McGraw-Hill Education

SCIENCE WORKBOOK FOR THE GED® TEST

SECOND EDITION

McGraw-Hill Education

SCIENCE WORKBOOK FOR THE GED® TEST

SECOND EDITION

McGraw-Hill Education Editors

Contributor: Jouve North America

New York Chicago San Francisco Athens London Madrid
Mexico City Milan New Delhi Singapore Sydney Toronto

1 2 3 4 5 6 7 8 9 LHS 23 22 21 20 19 18

ISBN: 978-1-260-12161-2
MHID: 1-260-12161-5

e-ISBN: 978-1-260-12162-9
e-MHID: 1-260-12162-3

Contents

Introduction

How to Use This Workbook

This workbook contains practice problems to help you test your science knowledge and reasoning skills in preparation for taking the GED® Science test.

Start your science practice by taking the **Science pretest** at the beginning of this workbook. It will help you decide which chapters of the workbook will be most valuable to you. Take the pretest in a controlled environment, with as few distractions as possible. If you want to closely simulate testing conditions, limit yourself to 90 minutes. When you are done, or when time is up, check your answers in the Answers and Explanations section at the end of the pretest. Next, find the problem numbers you answered incorrectly in the Evaluation Chart to identify the chapters on which you need to concentrate.

Each of the **three chapters** in the book has dozens of questions on one of the three content topics that are part of the GED® Science test. The questions have also been carefully designed to match each of the following:

- the test content
- the depth of knowledge (DOK) levels that are used to measure how well you understand each topic
- the Common Core State Standards (CCSS) that you are expected to have mastered

The exercises are not intended to be timed, but if you find that you are familiar with a topic, you could try timing yourself on a few problems, attempting to correctly work 5 questions in 10 minutes, for example. Answers for the exercises are located at the back of the workbook.

Finally, when you have completed the last exercise, take the **Science posttest** at the back of this workbook. This test can help you reevaluate yourself after practicing as much of the workbook as you feel is necessary. Answers are located at the end of the test, and another evaluation chart is provided to help you decide if you are ready to take the GED® Science test or where you might need further practice.

The GED® Science Test

The GED® Science test is a computer-based test, which allows for a broad range of item types. There are many multiple-choice items; each has four answer choices from which to choose. There are also many technology-based items with formats such as fill-in-the-blank, drop-down, and drag-and-drop.

- **Fill-in-the-blank:** These are short-answer items in which a response may be entered directly from the keyboard or in which an expression, equation, or inequality may be entered using an on-screen character selector with mathematical symbols not found on the keyboard.
- **Drop-down:** A list of possible responses is displayed when the response area is clicked with the mouse. These may occur more than once in a sentence or question.
- **Drag-and-drop:** Words are moved around the screen by pointing at them with the mouse, holding the mouse button down, and then releasing the button when the element is positioned over an area on the screen. Such items are used for sorting, classifying, or ordering questions.

About 40 percent of the questions on the GED® Science test focus on topics in the life sciences. Another 40 percent of the questions are on topics in the physical sciences. Twenty percent of the questions address topics in earth and space science. Many questions require you to read and interpret a document. That document might be text, a chart, a graph, a diagram, or a map.

Visit http://www.ged.com for more about the GED® test.

The Top 25 Things You Need to Know for the GED® Science Test

Use this list as a guide for your studies. Be sure to study and practice each topic until you feel that you have mastered it.

1. **DNA:** Know the structure of DNA and its function.
2. **11 Body Systems:** Know the functions and major organs of the 11 body systems (circulatory, lymphatic, digestive, endocrine, integumentary, muscular, nervous, reproductive, respiratory, skeletal, and urinary). Know how these systems interact.
3. **Cell Division:** Understand the processes of mitosis and meiosis.
4. **Photosynthesis:** Understand the process of photosynthesis.
5. **Cellular Respiration:** Understand the processes of respiration and fermentation.
6. **Ecosystem Disruptions:** Understand carrying capacity and the events that can disrupt an ecosystem.
7. **Feeding Levels:** Know the four feeding levels: producer, primary consumer, secondary consumer, and decomposer.
8. **Energy Flow:** Understand how several energy flow models work (food chain, food web, and pyramid).
9. **Water Cycle:** Understand the phases of the water cycle.
10. **Carbon and Oxygen Cycles:** Understand the phases of the carbon cycle and the oxygen cycle.
11. **Nitrogen Cycle:** Understand the phases of the nitrogen cycle.
12. **Heredity:** Know the difference between genotype and phenotype and how dominance works.
13. **Punnett Square:** Construct a Punnett square to predict ratios of inherited traits.
14. **Natural Selection:** Know the principles of natural selection.
15. **Solar System:** Know the planets in our solar system and their basic characteristics.
16. **Stars:** Understand the life cycle of stars.
17. **Plate Tectonics:** Understand the forces that shift the Earth's crust and the effects of those forces.
18. **Earth's Layers:** Describe the components of the four layers of the Earth (inner core, outer core, mantle, and crust), and define lithosphere and asthenosphere.

19. **Climate:** Understand the process of climate change and the effects humans have had on it.
20. **Atoms:** Know the structure of an atom and how atoms bond together.
21. **Chemical Reactions:** Understand chemical reactions and balance chemical equations.
22. **Motion:** Understand Newton's laws of motion and the functions of simple machines.
23. **Heat Transfer:** Understand the three ways heat can be transferred: conduction, convection, and radiation.
24. **Energy:** Understand conservation of energy, and know the various types of potential and kinetic energy.
25. **Waves:** Understand how waves work, and know the main types of waves.

McGraw-Hill Education

SCIENCE WORKBOOK FOR THE GED® TEST

SECOND EDITION

PRETEST

Science

40 questions | **90 minutes**

This pretest is intended to give you an idea of the topics you need to study to pass the GED® Science Test and includes topics from the three science disciplines represented by the three chapters in this workbook: Life Science, Physical Science, and Earth and Space Science.

Try to answer every question in a quiet area so you are free from distractions and with enough time. The usual time allotted for the test is 80 minutes. Remember that it is more important to think about every question than it is to finish ahead of time.

The pretest will give you an opportunity to identify your strengths and weaknesses so that you can focus your studies. Answers and explanations can be found at the end of the pretest.

Directions: Answer the following questions. For multiple-choice questions, choose the best answer. For other questions, follow the directions preceding the question.

PRETEST

Use the following diagram of a plant cell to answer Question 1.

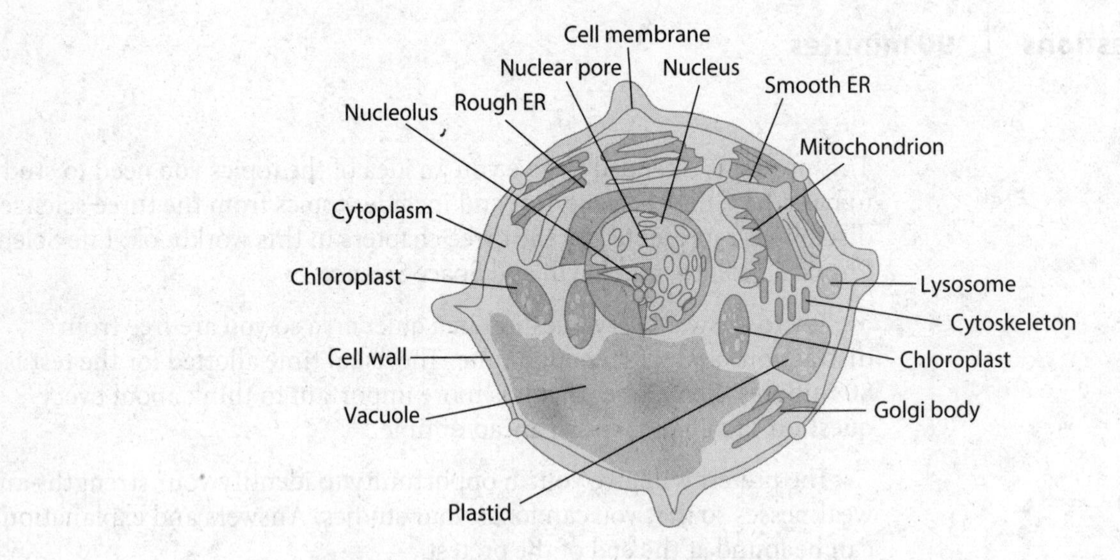

1. What is the function of the cell wall?

A. to build proteins
B. to convert solar energy into chemical energy
C. to protect and provide support for the cell
D. to take in carbon dioxide

PRETEST

Questions 2 and 3 are based on the following information.

The following reaction takes place in a reaction vessel:

$$\text{Heat} + A(s) + B(l) \rightarrow 2C(s) + D(s).$$

The chemist who is carrying out the reaction in the laboratory repeats the experiment under various conditions in order to produce the maximum possible amounts of substances C and D.

2. This reaction can be classified as

   A. endothermic
   B. exothermic
   C. at equilibrium
   D. a nuclear reaction

3. Which of the following factors will NOT help the chemist to change the speed or rate of the reaction?

   A. increasing the temperature of the system
   B. using powdered reactants
   C. increasing the pressure on the system
   D. adding a catalyst to the reaction

4. Which equation is NOT correctly balanced?

   A. $C + O_2 \rightarrow CO_2$
   B. $Na + Cl_2 \rightarrow 2NaCl$
   C. $Ca + Cl_2 \rightarrow CaCl_2$
   D. $3O_2 \rightarrow 2O_3$

PRETEST

Use the following information to answer Questions 5–9.

The Centers for Disease Control recommends that children in the United States receive certain vaccinations at given ages. Several of these vaccinations are repeated at specific intervals between birth and age six. Many of the diseases for which children are routinely vaccinated today used to pose serious health concerns in the past; however, since vaccinations began, some of these diseases have declined by 100 percent. For example, in the years just prior to the introduction of the measles vaccine, more than 503,000 cases were reported annually. In 2007, there were only 43 cases of the disease.

Recommended Immunization Schedule for Persons Aged 0 Through 6 Years—United States • 2011

For those who fall behind or start late, see the catch-up schedule

Vaccine ▼ Age ►	Birth	1 month	2 months	4 months	6 months	12 months	15 months	18 months	19–23 months	2–3 years	4–6 years
Hepatitis B[1]	HepB	HepB				HepB					
Rotavirus[2]			RV	RV	RV[2]						
Diphtheria, Tetanus, Pertussis[3]			DTaP	DTaP	DTaP	*see footnote[3]*	DTaP				DTaP
Haemophilus influenzae type b[4]			Hib	Hib	Hib[4]	Hib					
Pneumococcal[5]			PCV	PCV	PCV	PCV				PPSV	
Inactivated Poliovirus[6]			IPV	IPV		IPV					IPV
Influenza[7]								Influenza (Yearly)			
Measles, Mumps, Rubella[8]						MMR			*see footnote[8]*		MMR
Varicella[9]						Varicella			*see footnote[9]*		Varicella
Hepatitis A[10]							HepA (2 doses)			HepA Series	
Meningococcal[11]										MCV4	

Range of recommended ages for all children

Range of recommended ages for certain high-risk groups

PRETEST

5. According to the information given, which of the following can be most properly inferred?
   A. Children receive the varicella vaccine on the same schedule at which they receive the measles, mumps, and rubella vaccine.
   B. Children receive the rotavirus vaccine on the same schedule at which they receive the pneumococcal vaccine.
   C. All children receive the meningococcal vaccine between the ages of 2 and 6.
   D. The influenza vaccine is not given to children in high-risk groups.

6. According to the information given, which is a true statement?
   A. Children receive the rotavirus vaccine one time between ages 2 months and 6 months.
   B. All children should receive the meningococcal vaccine between the ages of 2 and 6 years old.
   C. Children are vaccinated against diphtheria, tetanus, and pertussis four times before their second birthday.
   D. The hepatitis B vaccine must be repeated at ages 6 months, 12 months, 15 months, and 18 months.

7. A patient receives the same vaccination each winter. According to the chart, which disease is this vaccination preventing?
   A. hepatitis A
   B. influenza
   C. pneumococcus
   D. varicella

PRETEST

8. Diphtheria, a disease caused by bacteria found in the mouth and throat, causes patients to suffer from a sore throat, as well as fever and chills. Untreated, it can lead to complications including heart failure and paralysis. It can be fatal in approximately 10 percent of those who contract the disease and used to be a major cause of death in children. As recently as the 1920s, about 15,000 people died from diphtheria each year. Which assumption can be made regarding this disease?

   A. More people die from diphtheria today than 100 years ago.
   B. The diphtheria vaccine has eliminated sore throats among young children.
   C. Diphtheria was the most serious health concern facing people in the 1920s.
   D. Widespread use of the diphtheria vaccine has significantly reduced its threat.

9. Pertussis, also known as whooping cough, looks like the common cold; however, after a few weeks, it causes patients to suffer violent coughing spells and may lead to pneumonia, seizures, and brain infections. In some cases, it can be fatal. It is spread through the air from one person to the next but can be prevented through vaccination. Which type of disease is pertussis?

   A. age-related
   B. environmental
   C. hereditary
   D. infectious

PRETEST

Use the following information to answer Question 10.

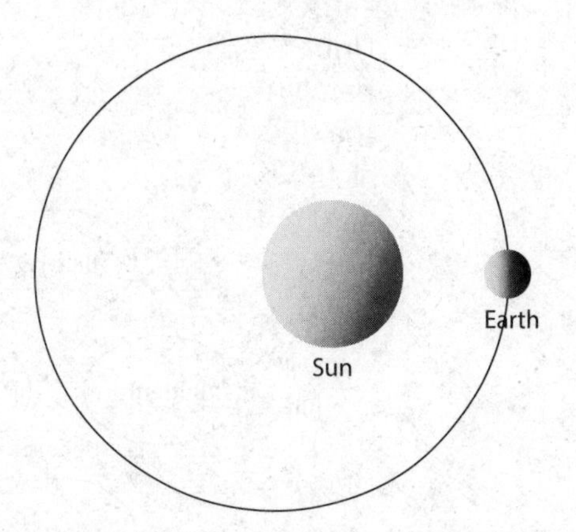

The following question contains a blank marked Select . . . ▼. *Beneath it is a set of choices. Indicate the choice that is correct and belongs in the blank. (**Note**: On the real GED test, the choices will appear as a "drop-down" menu. When you click on a choice, it will appear in the blank.)*

10. Earth travels around the sun in Select . . . ▼ orbit.
    - an elliptical
    - a circular

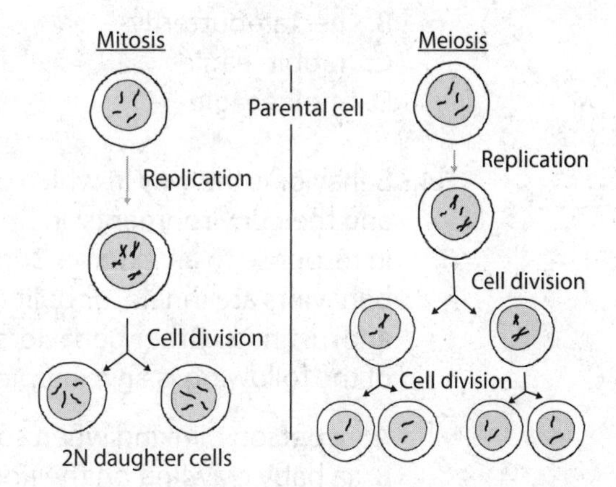

11. When comparing mitosis and meiosis, you can say that both processes

    A. produce the same number of daughter cells
    B. produce sex cells with the exact same number of chromosomes as the parent cells
    C. involve the replication of DNA
    D. will have mutations

PRETEST

Use the information that follows to answer Questions 12 and 13.

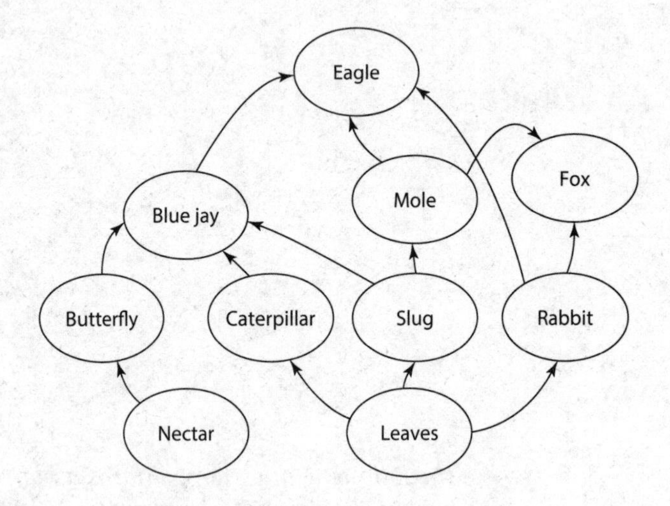

12. What is the role of the mole in the food web shown?

    A. secondary consumer
    B. decomposer
    C. tertiary consumer
    D. primary consumer

13. During which link in the food web is the most energy transferred?

    A. caterpillar–blue jay
    B. nectar–butterfly
    C. rabbit–eagle
    D. mole–eagle

14. Behavior is the way in which organisms interact with other organisms and their environments. In both people and animals, behavior occurs in response to an external stimulus, an internal stimulus, or both. Some behaviors are innate, or built in. These innate behaviors include reflexes and instincts. Other behaviors are learned as a result of experiences. Which of the following is an example of a reflex?

    A. a person blinking when something is thrown toward him or her
    B. a baby crawling on the floor before it begins to walk
    C. a bird gathering material to build its nest
    D. a dog barking when it hears a doorbell

PRETEST

Use the following diagram to answer Question 15.

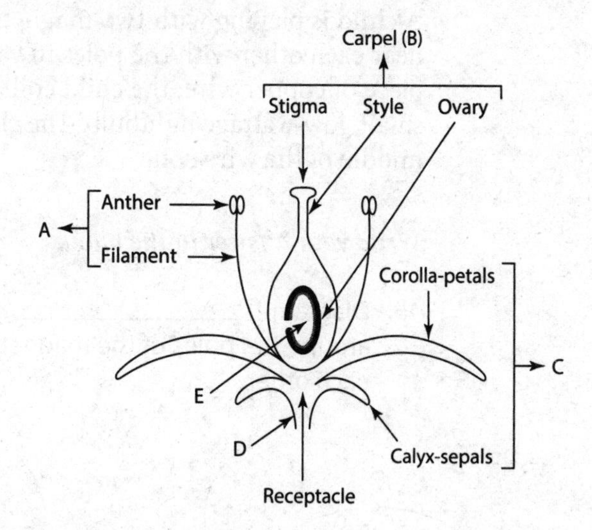

15. The anther and filament comprise which part of the flower?

A. ovule
B. pedicel
C. stamen
D. pistil

PRETEST

Questions 16 and 17 are based on the following passage.

A child is playing with two magnetic bars. First, the child places the bars near each other with the poles in various positions. Then, taking a long piece of copper wire, the child coils it and attaches the ends of the wire to a small, low-wattage lightbulb. The child then passes the magnets through the middle of the wire coil.

Write your answer in the blank.

16. Diagram ________________ could illustrate the field lines created around the poles of the magnetic bars when they are placed near each other.

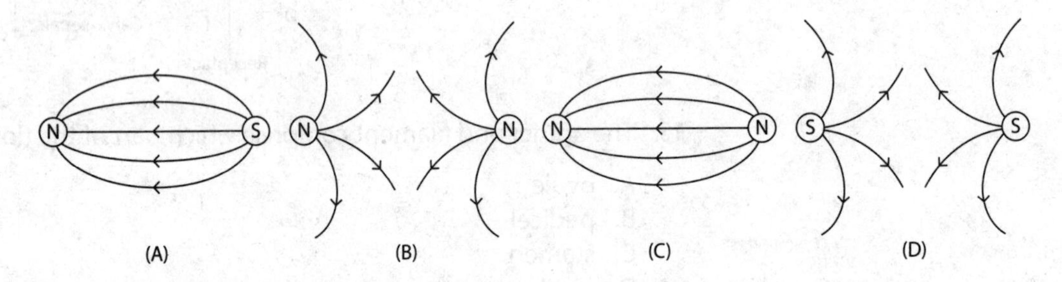

17. What happens when the magnetic bars are passed through the middle of the wire coil?

    A. The south poles of the magnets attract each other.
    B. An electromagnet is generated.
    C. The lightbulb lights up.
    D. The north and south poles of the magnets repel each other.

PRETEST

Questions 18 and 19 are based on the following passage.

A physicist is conducting an experiment regarding the momentum of objects in motion. She has a car, a baseball, a snowball, and a hockey player on ice skates ready to move. Each of the objects is set in motion, and its velocity is measured. The momentum of each object is then calculated. Next, the moving objects are set, two at a time, to collide with each other head on.

18. Which of the following has the greatest momentum?

    A. a 1,100-kg car moving at 20 m/s
    B. a 0.145-kg baseball moving at 35 m/s
    C. a 0.3-kg snowball moving at 15 m/s
    D. a 102-kg hockey player moving at 10 m/s

19. Which pair of objects will demonstrate conservation of momentum when they collide with each other?

    A. the hockey player and the baseball only
    B. the car and the baseball only
    C. none of the objects
    D. all of the objects

Questions 20 and 21 are based upon the following passage.

For the school science fair, Evelyn wanted to determine the optimal amount of water for marigold plants. She planted the same number of marigold seeds in each of 30 pots. She divided the pots into three groups, with 10 in each group, and labeled the groups A, B, and C. She watered the plants in group A daily, group B twice a week, and group C once a week. The plants received an equal amount of water each time. She predicted that the plants watered daily would be the tallest at the end of one month.

20. Which of the following would NOT be necessary for the experiment to be valid?

    A. using the same size pots
    B. including a control group
    C. planting all of the seeds in the same type of soil
    D. making sure the plants received an equal amount of sunlight

21. The watering schedule Evelyn used in the experiment is the

    A. dependent variable
    B. hypothesis
    C. independent variable
    D. result

PRETEST

Questions 22 and 23 are based on the following passage.

Laura accidently dropped a book that she had been holding up in the air. The book weighed 2.0 kg. It fell 1.8 m to the ground, making a loud noise when it hit. Assume that the acceleration due to gravity is 9.81 m/s^2 $P_E = mgh$.

22. How much potential energy did the book possess while in Laura's hands?

 A. 35.5 J
 B. 6.0 J
 C. 13.6 J
 D. 2.73 J

23. As the book fell through the air and eventually hit the ground,

 A. potential energy was destroyed and kinetic energy was created
 B. sound energy was created
 C. both potential energy and kinetic energy were destroyed
 D. the energies of the system were converted from one form to another

PRETEST

Questions 24 and 25 are based upon the following passage.

The physical structure of Earth includes three basic layers. At the center of the planet is a solid, very dense inner core surrounded by a liquid outer core. The outer core, which contains iron, spins as the planet rotates, generating a magnetic field as it flows. Together, the inner and outer cores are approximately 2,200 miles thick and make up about one-third of Earth's mass.

The core is covered by a layer called the mantle. It is semisolid, and although it is not the hottest layer of the planet, it is still so hot that some of the rock in this layer is molten. This rock flows slowly, similar to hot asphalt.

The mantle is covered by a rigid outside layer, or crust. This layer is about 25 miles thick beneath the continents and 4 miles thick beneath the oceans, which is relatively thin compared to the other two layers. The uppermost part of the mantle is cooler than the deeper parts and combines with the planet's thin crust to form the lithosphere. This layer has broken into pieces known as tectonic plates, which are constantly moving as they float on a layer of melted rock.

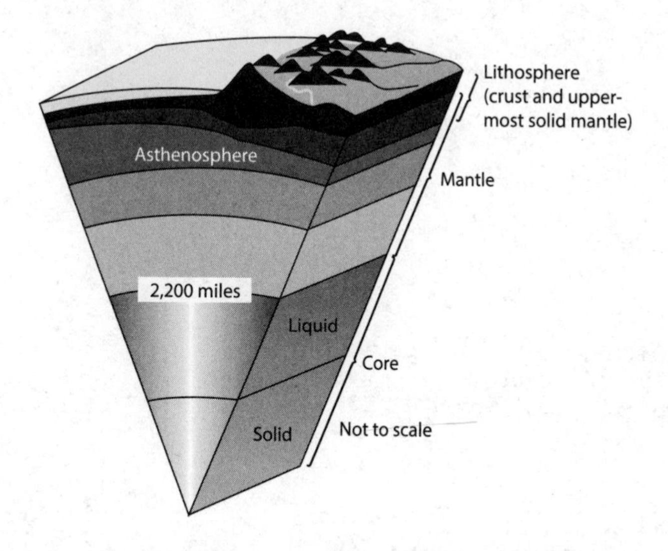

24. Which layer of Earth is responsible for a compass pointing toward north?

    A. crust
    B. outer core
    C. lithosphere
    D. mantle

25. Beneath which location is Earth's crust the thickest?

    A. Mount Everest
    B. Atlantic Ocean
    C. Grand Canyon
    D. Mojave Desert

PRETEST

26. Indicate the box in which each of the following items belongs.

- Cell wall
- Irregular shape
- Chloroplasts
- Plastids
- One or more small vacuoles

Plant cell	Animal cell

PRETEST

Questions 27 and 28 are based on the following passage.

Henri stops at a gas station and buys a cup of coffee that he heats in a microwave oven. He then proceeds on his way, enjoying music on the radio in his car. Some miles down the road, he is involved in an accident. Paramedics take him to a nearby hospital for a CT scan.

27. What type of radiation will be used during the CT scan?

    A. radio waves
    B. microwaves
    C. X-rays
    D. infrared radiation

28. Of all of the types of radiation mentioned in the passage, which one causes the LEAST amount of damage to living tissue and DNA with long-term exposure?

    A. All types of radiation are equally safe.
    B. radio waves
    C. microwaves
    D. the radiation used in the CT scan

29. While playing hockey, Brie is hit in the leg by a puck. The trainer immediately takes an ice pack at 0°C and places it on her injured leg, which is at 37°C. This incident occurs in an arena with an air temperature of 14°C. Which best explains the direction of heat flow in this system?

    A. The ice pack transfers coldness to the leg.
    B. The only direction of heat flow is from the leg to the ice pack.
    C. The heat from the leg is transferred to the air and to the ice pack, while heat from the air is transferred to the ice pack.
    D. Heat from the air is transferred to the leg.

PRETEST

Use the following diagram to answer Questions 30 and 31.

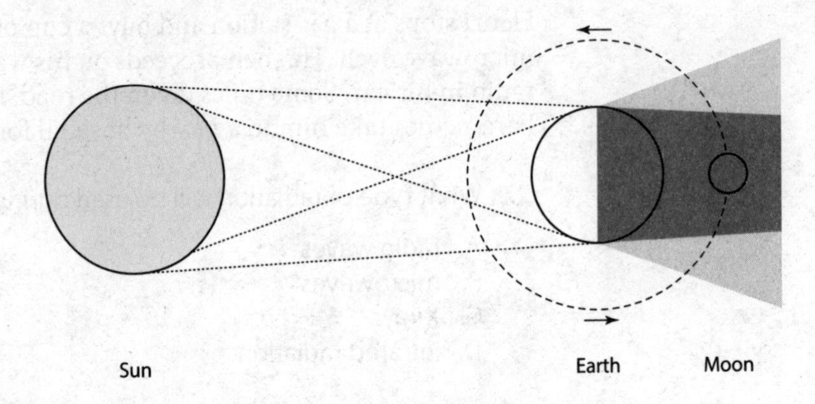

30. Which of the following does the diagram show?

    A. first quarter
    B. lunar eclipse
    C. new moon
    D. solar eclipse

31. Which of the following statements is an opinion rather than a fact?

    A. The length of time an eclipse is visible can vary.
    B. The phases of the moon occur in a repeated cycle.
    C. Certain types of tides are associated with the moon's phases.
    D. A lunar eclipse is more interesting to view than a solar eclipse.

PRETEST

Use the following diagram to answer Question 32.

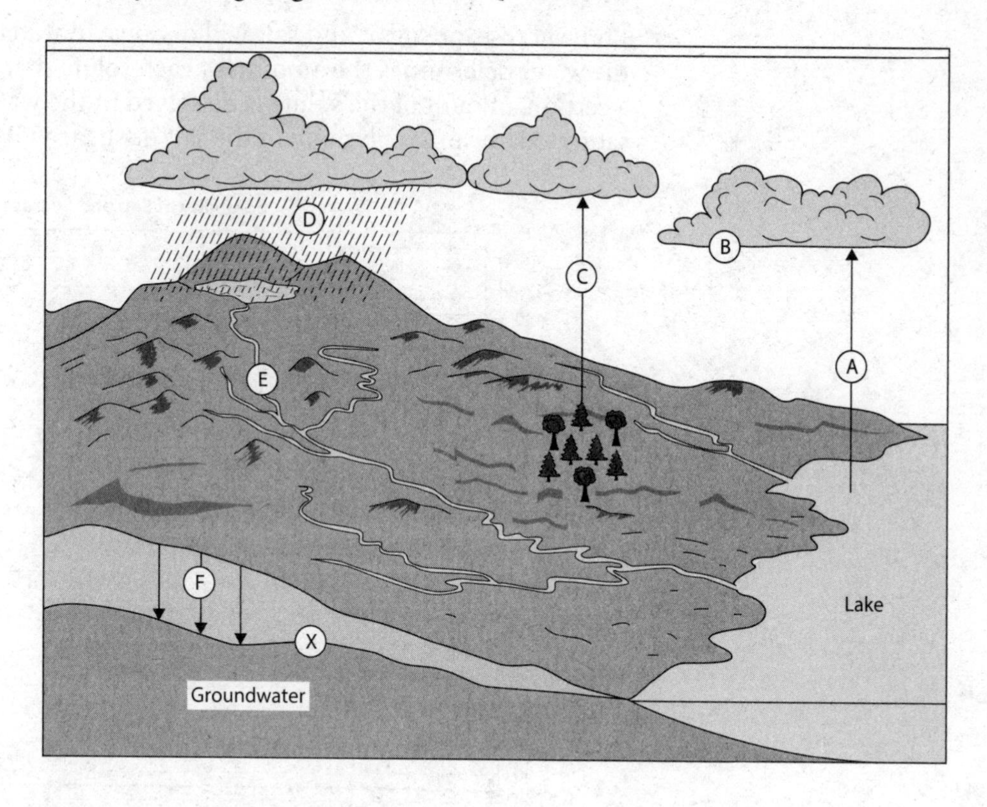

Write your answer in the blank.

32. In the hydrologic cycle shown above, letter A represents the process of

______________.

PRETEST

Use the following information to answer Questions 33–35.

Both sucrose, or sugar, and salt will dissolve in water. The temperature of the water determines the amount of each solute that can be dissolved. Once a certain amount of the solute is dissolved in the water, the water becomes saturated, meaning no more of the salt or sugar will dissolve in the solution.

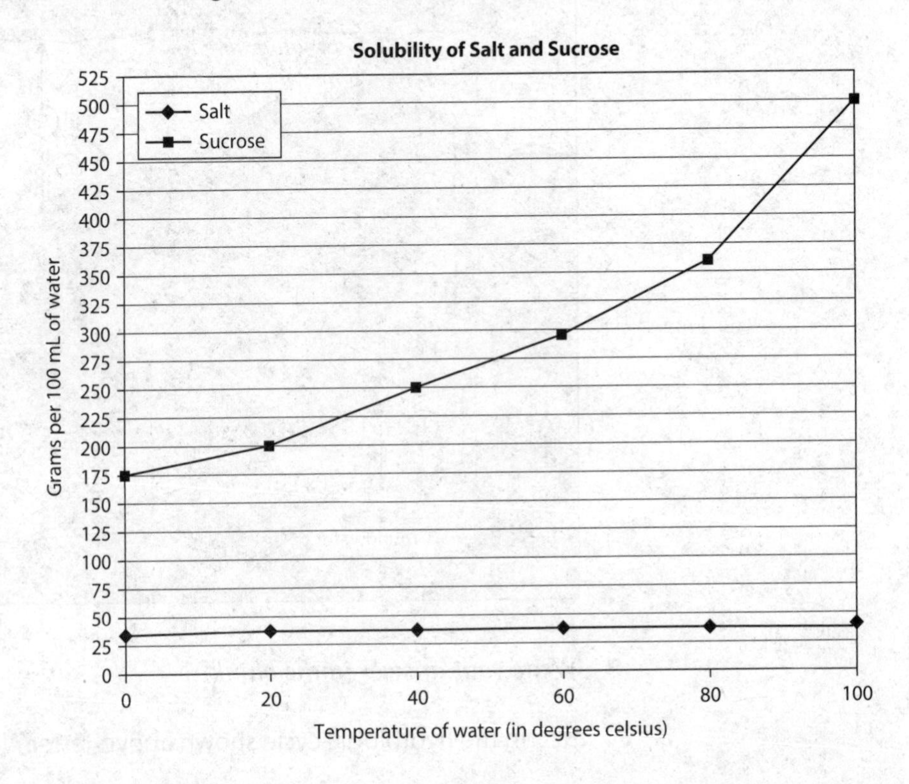

33. Approximately how many grams of sugar will saturate 100 mL of water at 40°C?

    A. 45 g
    B. 100 g
    C. 175 g
    D. 250 g

34. Which conclusion can be drawn regarding the data?

    A. When the temperature of the water increases, the amount of sucrose that can be dissolved decreases.
    B. Increasing the temperature of the water significantly increases the amount of salt that can be dissolved.
    C. When the temperature of the water increases, the amount of sucrose that can be dissolved increases as well.
    D. Increasing the temperature of the water by 100 percent also increases the amount of salt that can be dissolved by 100 percent.

PRETEST

35. Which of the following can be most properly inferred?

    A. The amount of sucrose that will dissolve in boiling water is approximately 10 times greater than the amount of salt that will dissolve.
    B. All solutes have an equal saturation rate when the temperature of the water reaches a certain point.
    C. At any temperature, the amount of sucrose that will dissolve in 100 mL of water is nearly 10 times greater than the amount of salt that will dissolve.
    D. The amount of sucrose that will dissolve in cold water is approximately the same as the amount of salt that will dissolve in the same water.

Use the following diagram to answer Question 36.

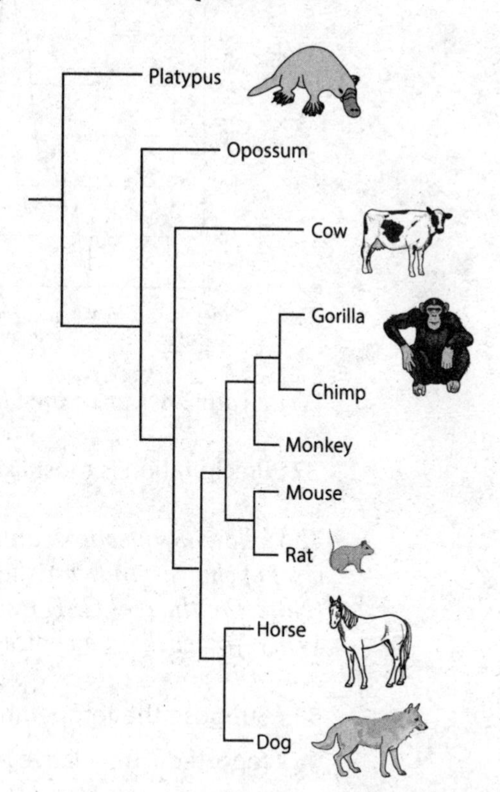

36. Which of the following pairs of animals is most closely related?

    A. opossum and platypus
    B. rat and mouse
    C. cow and horse
    D. monkey and gorilla

PRETEST

Use the following information to answer Questions 37 and 38.

Relative humidity indicates how much moisture is in the air compared to how much moisture the air can hold at that temperature. Generally, the amount of moisture in the air is less than the amount needed to saturate the air. When the air is saturated, the relative humidity will be near 100 percent.

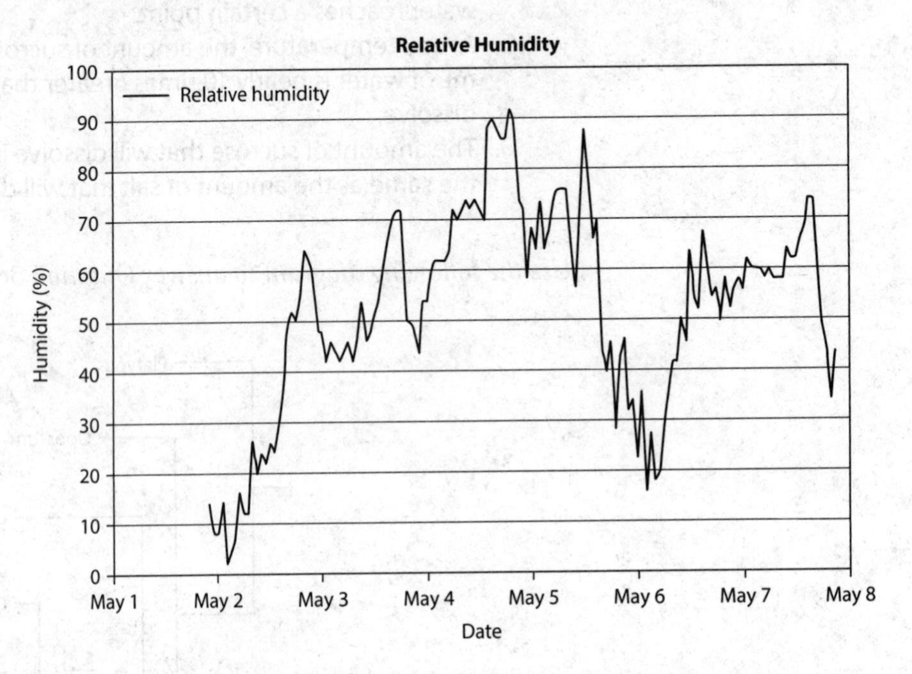

Write your answer in the blank.

37. Precipitation is most likely to occur on ____________________.

The following question contains a blank marked Select . . . ▼. *Beneath it is a set of choices. Indicate the choice that is correct and belongs in the blank. (**Note**: On the real GED test, the choices will appear as a "drop-down" menu. When you click on a choice, it will appear in the blank.)*

38. Suppose the temperature was 15 degrees higher on May 8 than originally reported. The relative humidity would Select . . . ▼.

decrease
increase
not change

PRETEST

Use the following information to answer Questions 39 and 40.

Levers, inclined planes, and pulleys are all examples of simple machines. A lever pivots around a point and is used to move or lift an object by applying force to the opposite end.

39. Which of the following is NOT an example of a lever?

    A. screw
    B. fork
    C. scissors
    D. crowbar

40. Which statement is true regarding the lever shown below?

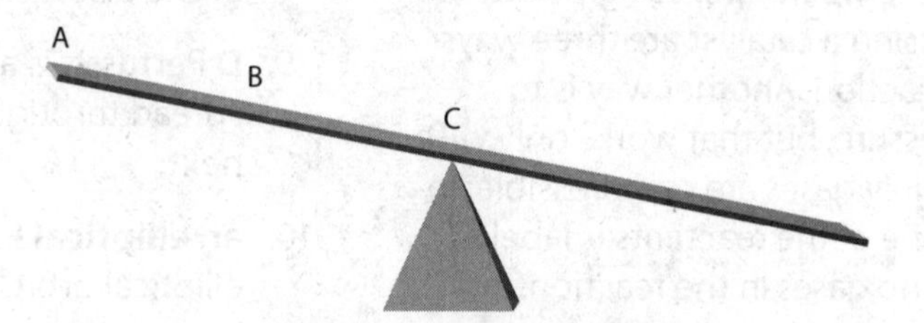

    A. The force in the diagram is applied to point B.
    B. The fulcrum in the diagram is located at point A.
    C. The fulcrum in the diagram is located at point C.
    D. The force in the diagram is applied to point A.

THIS IS THE END OF THE SCIENCE PRETEST.
ANSWERS AND EXPLANATIONS BEGIN ON THE NEXT PAGE

PRETEST

Answers and Explanations

1. **C** Even if you do not have previous knowledge of the function of the cell wall, you can look at the diagram and see that it completely surrounds the outside of the cell like a shell. A wall is usually for protection, and that is the case with a cell wall too.

2. **A** This reaction is endothermic because heat energy is being added to the system as a reactant. If the heat were a product, the reaction would be an exothermic reaction.

3. **C** Increasing the temperature, using powdered reactants, and using a catalyst are three ways to speed up a reaction. Another way is to increase the pressure, but that works only with gases because only gases are compressible. In the reaction, none of the reactants is labeled (g), so there are no gases in the reaction. Because there are no gases present, pressure cannot be used to speed up the reaction.

4. **B** According to the Law of Conservation of Mass, what is on the left side of the arrow must add up to what is on the right side. Choice A has one carbon atom and two oxygen atoms on both sides. Choice C has one calcium ion and two chlorine atoms/ions on both sides. Choice D has six oxygen atoms on both sides. Choice B is not balanced. The number of chlorine atoms on both sides is the same, but the number of sodium atoms/ions is not the same on both sides.

5. **A** Children receive the varicella vaccine on the same schedule at which they receive the measles, mumps, and rubella vaccine. The doses are given when children are between 12 and 15 months old and again between the ages of 4 and 6 years.

6. **C** It can be concluded from the chart that children are vaccinated against diphtheria, tetanus, and pertussis four times before their second birthday: at 2 months, 4 months, 6 months, and finally between 15 and 18 months.

7. **B** An influenza vaccination is recommended each winter.

8. **D** Since the paragraph says that until the 1920s, diphtheria was a deadly disease, one can assume that it is no longer such a threat. Putting the diphtheria vaccine on the recommended schedule for childhood vaccines has significantly reduced incidences of the disease.

9. **D** Pertussis is an infectious disease since it is spread through the air from one person to the next.

10. **an elliptical** Earth travels around the sun in an elliptical orbit.

11. **C** While meiosis produces sex cells with half as many chromosomes as the parent cell, mitosis produces cells with the exact same number of chromosomes as the parent cell. However, both processes involve the replication of DNA so that the cells produced during the cellular divisions will have an exact copy of the parent cell's DNA as needed.

12. **A** The mole is a secondary consumer because it eats slugs (primary consumers) that have eaten leaves.

13. **B** The most energy is transferred between the nectar and the butterfly. The greatest energy transfer in a food web is between producer and primary consumer. Nectar–butterfly is the only such relationship in the answer choices.

14. **A** Blinking when something is thrown toward you is an involuntary reflex action.

15. **C** The anther and filament are parts of the stamen of a flower.

16. **B** The field lines around a magnet move away from the north pole and toward the south

PRETEST

pole. Only diagram B shows the field lines moving away from the north poles.

17. **C** The movement of a magnet around a wire will cause an electric current to be produced. This current will flow through the wire and into the lightbulb, causing it to light up.
18. **A** To find the momentum of an object, multiply the mass of the object by its velocity. The car has the greatest momentum.
19. **D** Momentum is always conserved when objects collide, regardless of their masses or velocities.
20. **B** To test one variable, everything else must be the same. Since three different watering schedules are being tested, a control group is not necessary.
21. **C** An experiment attempts to answer a question. Variables are things that change. Evelyn is changing the watering schedule, so that is a variable. A dependent variable is the output or effect of the experiment, and an independent variable is the input or cause. Evelyn is changing the input by varying the watering schedule to see what the effects would be.
22. **A** The potential energy is measured by the equation $P_E = mgh$. Multiplying (2.0 kg)(9.81 m/s^2)(1.8 m) gives 35.5 J of energy.
23. **D** Energy is conserved; that is, it cannot be created or destroyed. Energy can be converted from one form to another. In this case the potential energy became kinetic energy, which then became sound energy.
24. **B** The lithosphere is the upper mantle and crust. A compass relies on magnetism to indicate direction. The passage states that the outer core generates a magnetic field, so a compass points north due to the magnetic field generated by the outer core.
25. **A** The crust is thinnest under the ocean and thickest under tall mountains. Of these answer choices, Earth's crust is thickest under Mount Everest.
26. Plant cell: cell wall, chloroplasts, plastids. Plant cells do have vacuoles, but they are large and there is usually just one. Animal cell: irregular shape, one or more small vacuoles.
27. **C** Of the choices, X-rays are the most powerful type of radiation and are used for medical purposes when needed. The other types of radiation listed are much weaker and are reserved for everyday use in the home.
28. **B** Radio waves are the weakest type of radiation mentioned in the passage. The CT scan uses X-rays, which are harmful with long-term exposure. Microwaves carry more energy than radio waves and can be dangerous to a person with a pacemaker.
29. **C** Heat will flow from a higher temperature to a lower temperature. Because the player's leg has the highest temperature, heat will flow from the leg to the ice pack and to the air in the arena. Because the air in the arena has a higher temperature than the ice pack, heat in the air will flow to the ice pack as well.
30. **B** The diagram shows a lunar eclipse in which the moon is completely within the shadow of Earth.
31. **D** Whether something is interesting is an example of an opinion rather than a fact.
32. **evaporation** Letter A shows evaporation of water from the lake into the atmosphere.
33. **D** 250 g is marked directly above 40°C on the horizontal axis of the graph.
34. **C** You can see from the chart that as temperature increases, the amount of solute that will dissolve increases.
35. **A** Water boils at 100°C. At this temperature, the amount of salt that will dissolve in 100 mL of water is approximately 40 g and the amount of sucrose that will dissolve is 500 g, which is approximately 10 times greater than the amount of salt that will dissolve.

PRETEST

36. **B** The diagram shows that the rat and the mouse are closely related. The other pairs share more distant common ancestors.

37. **May 5** Precipitation is most likely to occur on May 5 when the humidity is the highest.

38. **decrease** When the temperature is higher, humidity will decrease.

39. **A** A crowbar, fork, and scissors all move by applying force to one end. A screw does not.

40. **C** The lever shown looks like a playground seesaw. The lighter person would be sitting at point A and the heavier person would be on the opposite end. The pivot point, or fulcrum, is point C.

PRETEST

Evaluation Chart

Circle the item number of each question that you missed. To the right of the item numbers, you will find the names of the chapters that cover the skills you need to solve the questions. More question numbers circled in any row means more attention is needed to sharpen those skills for the GED® test.

Item Numbers	Chapters
1, 5, 6, 7, 8, 9, 11, 12, 13, 15, 20, 21, 26, 32, 36	Life Science
2, 3, 4, 14, 16, 17, 18, 19, 22, 23, 27, 28, 29, 33, 34, 35, 39, 40	Physical Science
10, 24, 25, 30, 31, 37, 38	Earth and Space Science

If you find you need instruction before you are ready to practice your skills with this workbook, we offer several excellent options:

McGraw-Hill Education Preparation for the GED Test: This book contains a complete test preparation program with intensive review and practice for the topics tested on the GED.

McGraw-Hill Education Pre-GED: This book is a beginner's guide for students who need to develop a solid foundation or refresh basic skills before they embark on formal preparation for the GED test.

McGraw-Hill Education Short Course for the GED: This book provides a concise review of all the essential topics on the GED, with numerous additional practice questions.

CHAPTER 1

Life Science

Questions about life science make up about 40 percent of the GED Science test. Life science, as the name suggests, is the science of living organisms in the world around us. This portion of the GED Science test covers various topics, including the functions of the different body systems and how they are interrelated. It covers how diseases are spread by pathogens, our body's mechanism to fight diseases, and topics related to health and nutrition. Also included are topics about the relationships between life functions and the flow of energy—the process by which food is manufactured by green plants via photosynthesis and how energy from green plants reaches other organisms. Relationships between different organisms for survival and factors affecting population of organisms are also covered in this portion.

The life science chapter of this workbook familiarizes students with the organization of life—the cell theory of life, the mechanism of genetic inheritance, and the evidence of common ancestry. A brief idea of an organism's changes in response to the environment will also be covered. On completing this chapter, students will have gained valuable practice working the types of questions about living organisms and their interactions that will be on the GED Science test.

Directions: Answer the following questions. For multiple-choice questions, choose the best answer. For other questions, follow the directions preceding the question. Answers begin on page 163.

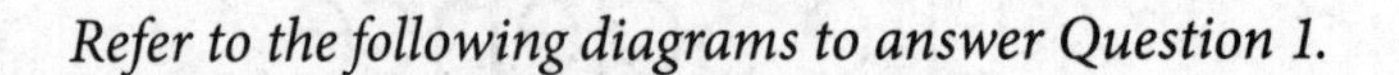

Refer to the following diagrams to answer Question 1.

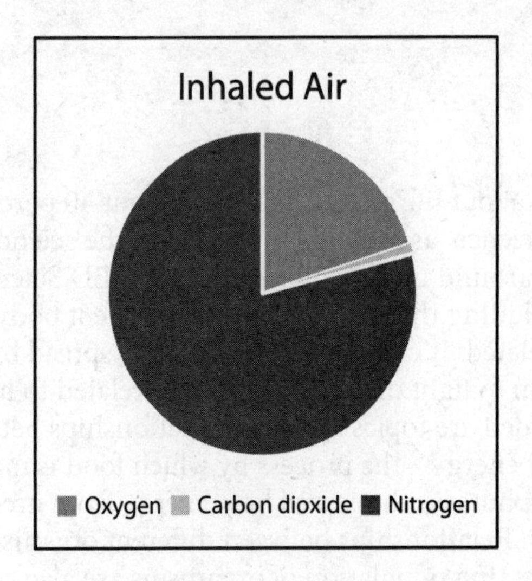

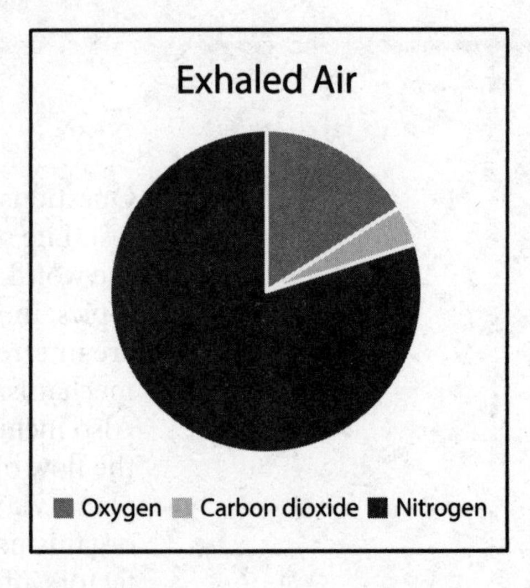

1. What fact about human respiration is confirmed from the circle charts?
   A. Oxygen is used and carbon dioxide is released during respiration.
   B. Nitrogen is a by-product of respiration.
   C. Oxygen, carbon dioxide, and nitrogen are the only three gases involved in respiration.
   D. Nitrogen is the most important gas for respiration.

2. Lime water turns milky when carbon dioxide is passed through it. Keeping this in mind, design a simple experiment to show that carbon dioxide is exhaled during respiration.

Write a short description of your experiment on the lines.

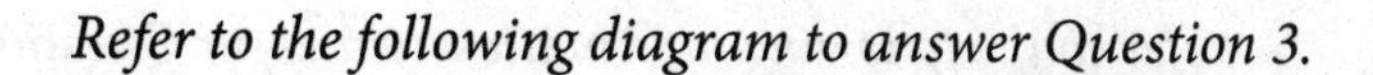

Refer to the following diagram to answer Question 3.

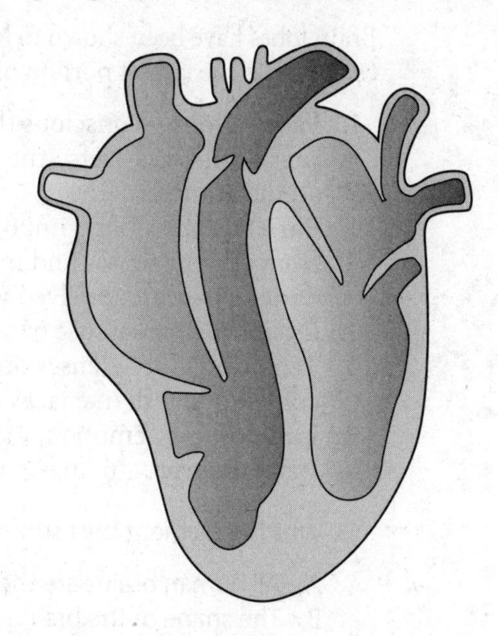

3. What is the main purpose of the organ pictured above?
   A. to absorb oxygen from air
   B. to digest food
   C. to filter blood
   D. to pump blood

4. What is the name given to the protective outer layer of a tooth?
   A. crown
   B. enamel
   C. neck
   D. root

5. In which of these body systems does the tooth play a significant role?
   A. circulatory system
   B. digestive system
   C. nervous system
   D. endocrine system

Refer to the following information to answer Questions 6 and 7.

Brain lobes have been shown to be related to different brain functions. The cerebrum is the largest portion of the human brain and is divided into six lobes.

1. **Frontal lobe**—conscious thought; damage can result in mood changes, social differences, etc.; the frontal lobes are the most uniquely human of all the brain structures
2. **Parietal lobe**—plays important roles in integrating sensory information from various senses and in the manipulation of objects; portions of the parietal lobe are involved with visuospatial processing
3. **Occipital lobe**—sense of sight; lesions can produce hallucinations
4. **Temporal lobe**—senses of smell and sound, as well as processing of complex stimuli like faces and scenes
5. **Limbic lobe**—emotion, memory
6. **Insular cortex**—pain, some other senses

6. Which statement best summarizes the information in the passage?

   A. All human brains are the same size.
   B. The shape of the brain varies from person to person.
   C. Specific parts of the human brain control specific body functions.
   D. The size of the brain is related to the organism's intelligence.

7. We remember the faces of people we saw a long time ago. Which lobes of the brain are most likely responsible for this?

   A. frontal and temporal
   B. parietal and occipital
   C. occipital and temporal
   D. frontal and parietal

Refer to the following information to answer Questions 8 and 9.

The two types of nerves that carry messages to and from the brain are the sensory and motor nerves. **Motor nerves** carry messages from the brain to control muscles and operate body systems. **Sensory nerves** carry messages from the surrounding environment to be processed by the brain.

8. When an ant bites you on the hand, what is most likely to happen?

   A. The hand informs the motor nerves that an ant is biting it, which in turn tells the sensory nerves to kill it.
   B. The sensory nerves on the hand send impulses to the brain about the ant. The brain signals the motor nerves to remove the ant.
   C. The motor nerves on the hand send impulses to the brain about the ant. The brain signals the sensory nerves to operate the hand to remove the ant.
   D. The brain signals both the motor and sensory nerves, and together they help the hand remove the ant.

Write your answer in the blank.

9. The optic nerves transmit visual information captured by the retina to the brain. They can be classified as ________________ nerves.

Refer to the following information to answer Questions 10–12.

The three major structural tissues in the human body are fat, muscle, and bone. Fat can be of two types: essential fat or storage fat. Essential fat is stored in bone marrow and in organs. Storage fat is the type we usually think of as excess fat, since it often adds unwanted weight to our bodies. Storage fat is fatty tissue that surrounds internal organs or is deposited beneath the skin.

The following circle graphs compare the body composition of an average-sized young man with that of an average-sized young woman.

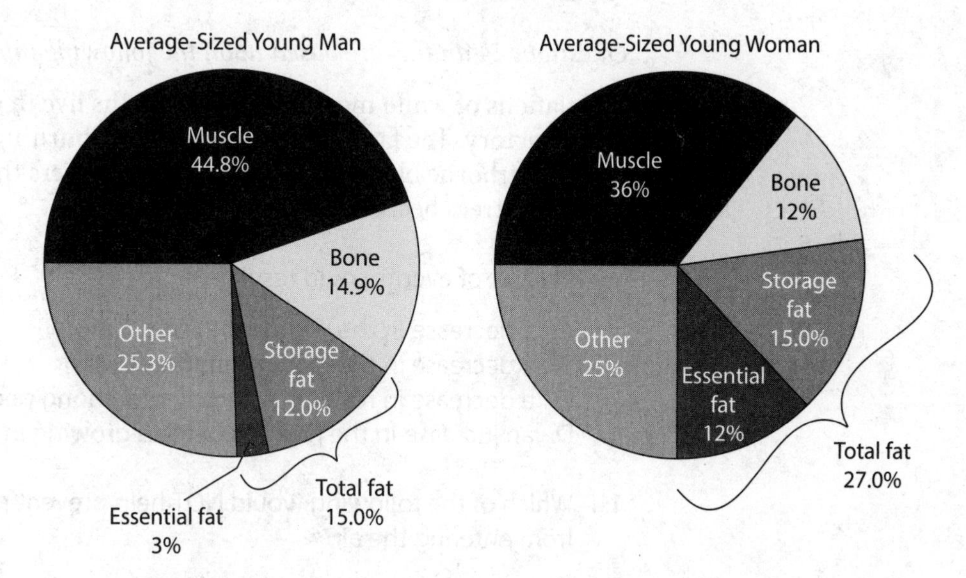

10. Which tissues make up a greater percentage of the body composition of an average-sized young man than of an average-sized young woman?

    A. total fat and bone
    B. bone and muscle
    C. total fat and muscle
    D. muscle only

11. Isabella is an average-sized young woman, and Noah is an average-sized young man. Which statement about Isabella and Noah is best supported by the passage and the graphs?

    A. Isabella has a more difficult time losing weight than does Noah.
    B. Isabella has a higher percentage of total fat than does Noah.
    C. Isabella has the same percentage of total fat as Noah does.
    D. Isabella eats more sweets than Noah does.

12. What is the most reasonable explanation for why women tend to have a higher percentage of total fat than men?

    A. Women have broader hips than men.
    B. Women have larger appetites than men.
    C. Women's bodies use added fat to protect female reproductive organs.
    D. Women tend to do less physical labor than men.

13. The phrase "survival of the fittest" best describes

    A. mitosis
    B. mutations
    C. natural selection
    D. recessive traits

Questions 14 and 15 are based upon the following information.

Populations of white moths and black moths live in a forest that is located near a factory. The factory is powered by the burning of fossil fuels, which releases airborne black soot that eventually covers the trees in the area. As a result, the trees become darker in color.

14. This set of events could result in

    A. a decrease in the white moth population
    B. a decrease in the black moth population
    C. a decrease in respiratory problems among people living nearby
    D. an increase in the number of trees growing in the area

15. Which of the following would NOT help prevent pollution from the factory from entering the air?

    A. using solar energy to power the factory
    B. using wind turbines to power the factory
    C. driving hybrid vehicles
    D. using steam to power the factory

Refer to the following information to answer Questions 16 and 17.

Arteries are blood vessels that carry blood away from the heart. Veins are blood vessels that carry blood toward the heart. In contrast to arteries, veins are less muscular and are often closer to the skin. Most veins (with the exceptions of the pulmonary and umbilical veins) carry deoxygenated blood from the tissues back to the heart. Most veins contain valves, which are bicuspid structures that look like two flaps made of elastic tissue. When the surrounding muscles contract, blood inside the veins is squeezed up the vein and the valves open. When the muscles relax, the valves close under the weight of the blood to prevent backflow.

16. What is the purpose of valves in veins?

    A. to increase the speed of blood flow
    B. to restrict the flow of blood to a single direction
    C. to enable blood to flow back and forth
    D. to transfer blood from arteries to veins

17. Which of these has an action similar to that of the valves in veins?

    A. a bicycle tire
    B. the filament of a lightbulb
    C. a garden hose
    D. the motor of a fan

Use the following information and diagram to answer Question 18.

An evolutionary tree is one way of showing evolution from a common ancestor. A sample evolutionary tree is shown below.

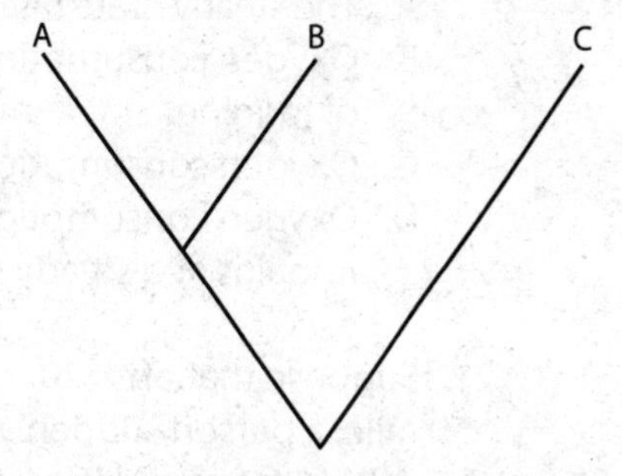

The points along the diagonal line where the links between species intersect indicate a common ancestor. The nearer the point of intersection, the more closely related are the two species.

The following question contains a blank marked Select... ▼. *Beneath it is a set of choices. Indicate the choice that is correct and belongs in the blank. (**Note:** On the real GED test, the choices will appear as a "drop-down" menu. When you click on a choice, it will appear in the blank.)*

18. Based on the diagram, species A and B are more Select... ▼ related than are species A and C.

Select... ▼
closely
distantly

Refer to the following information to answer Questions 19–22.

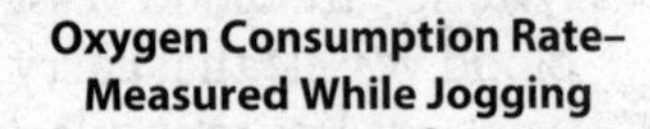

Oxygen Consumption Rate– Measured While Jogging

(measured at a slow-jog pace of 12 minutes per mile)

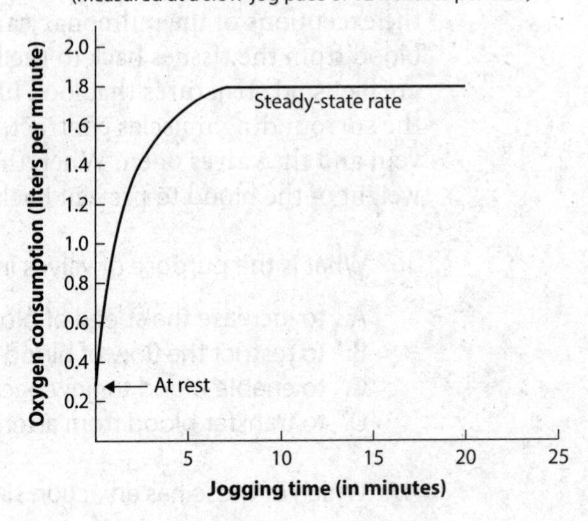

19. According to the graph, about how much oxygen is consumed per minute by a person at rest?

   A. 0.1 liter
   B. 0.3 liter
   C. 0.8 liter
   D. 1.0 liter

20. Which statement best summarizes the graph?

   A. The steady-state rate of oxygen consumption depends on jogging speed.
   B. Oxygen consumption increases most rapidly during the first 6 minutes of jogging.
   C. Oxygen consumption is lower while at rest than during jogging.
   D. Oxygen consumption rises during the first 6 minutes of jogging and remains at a steady state while jogging continues.

21. Suppose that after 10 minutes of slow jogging at a rate of 12 minutes per mile, a person suddenly speeds up to a rate of 10 minutes per mile. What will be the most likely effect on oxygen consumption?

   A. an increase to a higher steady-state level
   B. a steadily falling rate and no new steady-state level
   C. a continuation of the same steady-state level
   D. a decrease to a lower steady-state level

22. Which of the following might be the jogger's rate of oxygen consumption if she continues to run at 10 miles per hour for another 5 minutes?

   A. 1.5 liters per minute
   B. 1.7 liters per minute
   C. 1.8 liters per minute
   D. 2.0 liters per minute

Refer to the following diagram to answer Questions 23 and 24.

Comparison of Human and Baboon Jaws

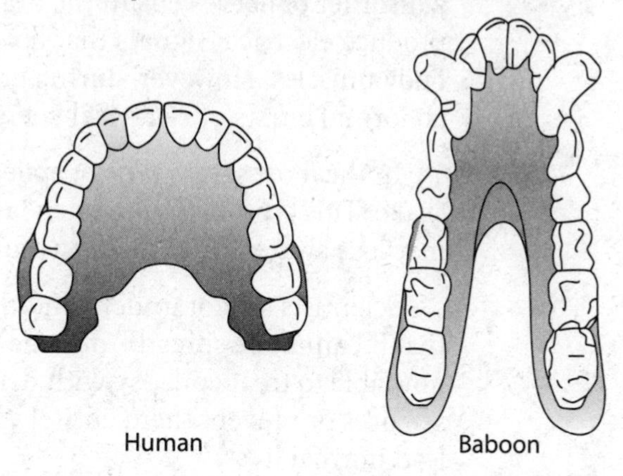

23. What feature of a human jaw is similar to that of a baboon?

    A. length of the jaw
    B. width of the jaw
    C. shape of the teeth in the jaw
    D. number of teeth in the jaw

24. Differences between the jaw of a human being and the jaw of a baboon most likely are related to differences in what activity?

    A. breathing
    B. diet
    C. intelligence
    D. life span

25. Lung tissue is damaged when exposed to smoke over a prolonged period of time. When lung tissue is damaged, less oxygen enters the bloodstream from the lungs. Suppose that a smoker and a nonsmoker begin an exercise program. Compared with the nonsmoker, what would the smoker most likely experience during mild exercise?

    A. a more rapidly beating heart
    B. less need to rest frequently
    C. muscles tiring more slowly
    D. a slower breathing rate

26. Which fact is LEAST relevant to the relationship between the consumption of alcohol and good health?

    A. Alcohol affects some body organs more than others.
    B. One can of beer contains the same amount of alcohol as one glass of wine.
    C. Seventy percent of drivers killed in one-car accidents had been drinking alcohol shortly before the accident.
    D. Alcohol intake by pregnant women may harm the fetus.

Refer to the following information to answer Questions 27 and 28.

Two million Americans suffer from some form of epilepsy. Epilepsy is a disorder of nerve cells in the brain. Normally, nerve cells in the brain produce electrical signals that flow through the nervous system and activate body muscles. However, during an epileptic seizure, these cells release abnormal bursts of electrical energy that the brain cannot control.

In the most severe type of epilepsy, people may lose consciousness, fall, and shake. This type of seizure often lasts several minutes. In a milder form, people may lose awareness of their surroundings but do not fall or lose consciousness.

Scientists do not understand the causes of epilepsy, but they do know that it cannot be spread from one person to another. Moreover, doctors are now able to treat epilepsy with drugs that either reduce the frequency of seizures or prevent them entirely. Most people who have epilepsy can now lead normal lives.

27. Which of the following is the best description of an epileptic seizure?

    A. a temporary loss of body control due to an electrical disturbance in the brain
    B. a permanent loss of body control due to an electrical disturbance in the brain
    C. an electrical disturbance in the brain caused by a loss of body control
    D. an injury to the body caused by a fall that occurs because of a brain malfunction

28. Which phrase is LEAST descriptive of epilepsy?

    A. electrical disturbance in the brain
    B. a contagious disease
    C. often accompanied by a seizure
    D. characterized by loss of body control

Refer to the following information to answer Questions 29 and 30.

Dr. Parkerson performs a simple test on her patient by holding a small ball and having her patient hold one hand beneath it. The doctor then releases the ball, and the patient attempts to grab the ball in his fist.

29. What is the experiment above most likely designed to test?

    A. hand strength
    B. strength of gravity
    C. weight of the ball
    D. reaction time

30. Which activity done prior to the test would most likely influence a person's ability to perform well on this test?

    A. eating vegetables
    B. drinking alcohol
    C. playing computer games
    D. talking on the phone

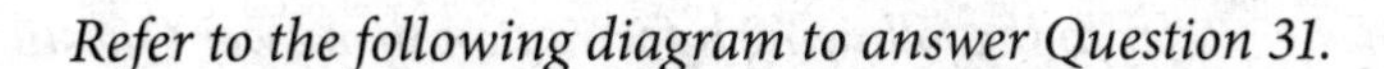

Refer to the following diagram to answer Question 31.

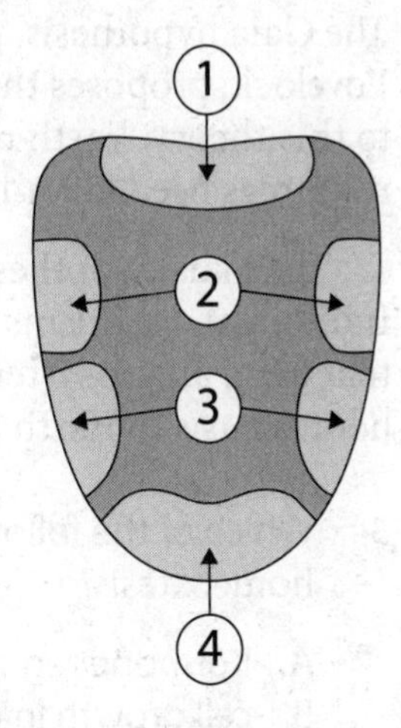

31. The tongue is the primary organ of taste. It consists of taste buds on its upper layer which helps it identify the tastes. Individual areas of the tongue are able to identify tastes separately. Bitter, sour, salty, and sweet are the tastes identified on the diagram by areas 1, 2, 3, and 4, respectively. Which portion of the tongue will taste a piece of candy?

    A. area 1
    B. area 2
    C. area 3
    D. area 4

Refer to the following information to answer Questions 32 and 33.

Homeostasis refers to the ability of cells or body systems to keep a stable internal environment as the external environment changes. Cells and body systems monitor their internal environment and adjust the internal conditions if a change occurs. These adjustments may be made through a negative feedback system or a positive feedback system.

A **negative feedback** system is a process in which an initial change will bring about an additional change in the opposite direction.

A **positive feedback** system is a process in which an initial change will bring about an additional change in the same direction.

Write your answer in the blank.

32. When skin is cut and blood vessels get damaged, platelets start clinging to the injured site and release chemicals that attract more platelets. The platelets continue piling up and releasing chemicals until a clot is formed.

    This is an example of a ____________________ feedback.

Write your answer in the blank.

33. When the blood sugar level in a person's blood rises, receptors in the body sense a change. The pancreas in turn secretes insulin into the bloodstream, which lowers the blood sugar level. This can be referred to

    as a ____________________ feedback.

Refer to the following information to answer Questions 34 and 35.

The Gaia hypothesis, proposed in the 1970s by British scientist James Lovelock, proposes that Earth behaves as a single living organism. According to this theory, Earth regulates its own temperature, provides itself with resources needed for life, disposes of its own wastes, and fights off disease.

The Gaia hypothesis is an extension of the concept of homeostasis. Individual organisms exhibit homeostasis, as do communities of organisms that form an ecosystem. The Gaia hypothesis applies the concept of homeostasis to Earth and views it as a single self-regulating system.

34. Which of the following is LEAST likely to be classified as an example of homeostasis?

    A. hormone regulation in plants
    B. cell growth in plants
    C. surrogate parenting
    D. warming effect of Earth's atmosphere

35. According to the Gaia hypothesis, which of the following would most likely harm Earth to the greatest degree?

    A. lightning-caused fires in the tropical rain forests
    B. several earthquakes along major fault lines on each continent
    C. collision with a large asteroid
    D. volcanic eruptions along the rim of the Pacific Ocean

Refer to the following information to answer Questions 36–39.

Heatstroke (sunstroke) occurs when the temperature-regulating system of the body ceases to work effectively. When people are exposed to or undergo heavy exertion in extreme heat, they may stop sweating—the body's natural way of cooling itself. The skin becomes hot and dry, and body temperature rises above normal. Other symptoms may include irregular heartbeat and shallow, irregular breathing. The victim may lose consciousness.

Because the high body temperatures of heatstroke can cause brain damage and death, heatstroke should be treated immediately. Standard treatment is to quickly reduce body temperature by applying cold compresses to the victim's body and ice packs to the neck. If possible, the victim should be placed in a bathtub of cold water. Only when body temperature is down to 102°F should the cooling procedures be stopped.

Heat exhaustion is less serious than heatstroke. With heat exhaustion, a victim becomes weak and dizzy after exposure to or heavy exertion in high temperature and high humidity. Other symptoms include confusion, a great amount of sweating, and a body temperature below normal. Proper treatment includes moving the victim to a cooler location but keeping the victim warm until body temperature rises to normal.

To help avoid heatstroke and heat exhaustion, people exposed to hot and possibly humid conditions should drink plenty of water and take frequent rest breaks to cool off.

36. Which of the following is NOT a symptom of heat exhaustion?

    A. dizziness
    B. profuse sweating
    C. elevated temperature
    D. confusion

37. What is the main purpose of applying ice packs to the neck of a heatstroke victim?

    A. to slow the flow of blood to the brain
    B. to lower body temperature
    C. to cool the nervous system
    D. to increase the flow of blood to the brain

38. To help prevent heat exhaustion during summer, which is the LEAST important for an exercise club to maintain in good working order?

    A. drinking fountains
    B. ventilation system
    C. air conditioner
    D. exercise bicycles

39. A doctor comes to the assistance of a middle-aged woman who has collapsed while working outdoors on a hot afternoon. Which of the following is the LEAST important information for the doctor?

    A. body temperature
    B. blood pressure
    C. air temperature
    D. breathing rate

Refer to the following information to answer Questions 40 and 41.

Recommended Daily Vitamin Needs of Male and Female Adults*

(ages 19–30)

	Males	*Females*
Vitamin A	1,000 μg	800 μg
Vitamin E	10 mg	8 mg
Vitamin K	75 μg	62 μg
Vitamin C	60 mg	1.1 mg
Thiamin	1.2 mg	1.1 mg
Riboflavin	1.3 mg	1.3 mg
Niacin	16 mg	14 mg
Vitamin B_6	1.3 mg	1.3 mg
Folate	400 μg	400 μg
Vitamin B_{12}	2.4 μg	2.4 μg
Vitamin D	5 μg	5 μg

mg = milligram; μg = microgram

Based on figures from the National Academy of Sciences

40. Which of the following can be inferred from the chart?

 A. Vitamin A is more necessary than all other vitamins.
 B. The human body needs a variety of vitamins for good health.
 C. Children do not need vitamins.
 D. Vitamin A is important for night vision.

41. Which of the following is an opinion rather than a scientific fact?

 A. Vitamin deficiency can lead to many types of illnesses.
 B. Fruits and nuts contain more essential vitamins than candy.
 C. Vitamin supplements should be taken by everyone over the age of 30.
 D. Both minerals and vitamins are essential for good health.

Based on the following information, answer Questions 42 and 43.

Pathogens broadly refer to anything that may cause a disease. Some types of diseases caused by pathogens include the following:

- **Air-borne diseases**—Any disease that is caused by pathogens transmitted through the air
- **Water-borne diseases**—Any disease that is transmitted by pathogenic microorganisms contaminated in fresh water
- **Food-borne diseases**—Any disease that is transmitted by pathogens present in food
- **Blood-borne diseases**—Any disease in which the pathogen is transmitted via the blood vessels

42. Which of these are likely to spread the following type of diseases? Write each of the following in the appropriate box below.
    - Sneezing
    - Infected syringe
    - Open wound
    - Dusting

Air-borne	Blood-borne

43. Which of the following is LEAST likely to prevent the spread of disease?
    A. using a tissue while sneezing
    B. washing hands before eating
    C. following safe cooking practices
    D. going to sleep early

Refer to the following information to answer Questions 44–48.

Salmonellosis is caused by rod-shaped bacteria called *Salmonella* that are present in moist, high-protein foods. These foods include meat, poultry, milk products, and egg products. *Salmonella* bacteria are often carried by flies, other insects, and household pets. *Salmonella* can usually be destroyed by cooking and by good hygiene. However, *Salmonella* can easily contaminate foods, even saved cooked foods, that are not refrigerated.

Food poisoning from *Salmonella* occurs most often during summer months and is often contracted at barbecues where food handling and hygiene may be careless. To reduce chances of *Salmonella* poisoning, refrigerate all perishable foods and make sure that people who handle food have clean hands.

44. Which of the following foods is LEAST likely to carry the *Salmonella* bacteria?

    A. salted crackers
    B. duck eggs
    C. oysters
    D. cottage cheese

45. During which month are the greatest number of cases of salmonellosis in the United States most likely to occur?

    A. April
    B. July
    C. October
    D. January

46. Which of the following is the LEAST likely source of *Salmonella*?

    A. a plate of raw hamburgers at a summer barbecue
    B. a picnic table containing scraps of food
    C. flies buzzing around garbage cans
    D. a campfire site

47. A family at an afternoon barbecue wants to be protected against food poisoning. Which of the following is the LEAST important action for the family to take?

    A. Toast the hamburger buns before making sandwiches.
    B. Thoroughly cook chicken and hamburger before serving.
    C. Clean the surface of the grill before cooking.
    D. Clean the picnic table before setting utensils down.

48. To prevent the possibility of food poisoning, which of the following should a shopper always check before buying a food item?

    A. the date on which it was produced
    B. the list of ingredients it contains
    C. the number of calories per serving
    D. the date after which it should not be sold

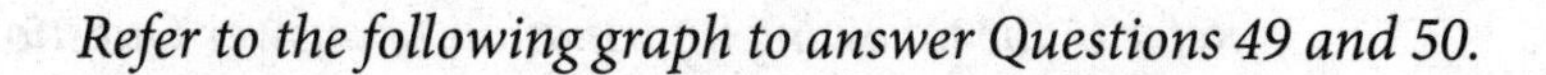

Refer to the following graph to answer Questions 49 and 50.

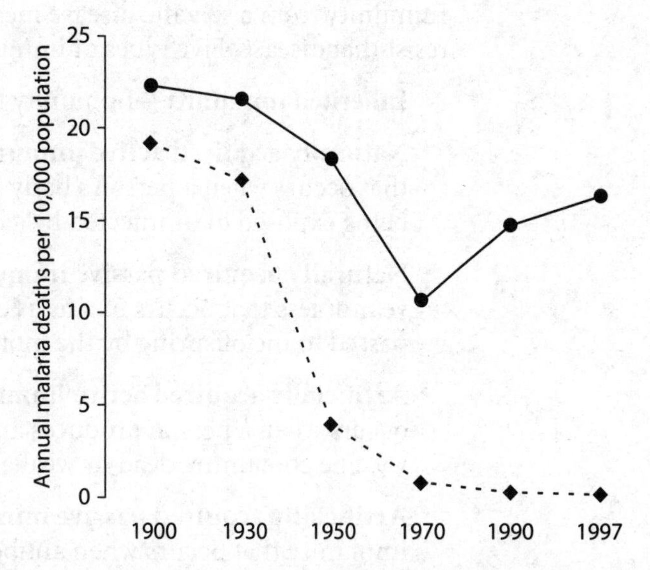

The graph represents the malaria death rate worldwide in the twentieth century. The top line includes sub-Saharan Africa. The bottom line excludes it.

49. Which of the following can be determined from the graph?

    A. The death rate in the 1990s was the lowest.
    B. The sub-Saharan death rate increased after 1970 even as the death rate in the rest of the world decreased.
    C. The world death rate excluding sub-Saharan death is greater than the death rate including it.
    D. There was a steep increase in the death rate between 1930 and 1970.

Write your answer in the blank.

50. The highest number of deaths from malaria worldwide was recorded in the year ________.

Refer to the following diagram to answer Question 51.

Atoms → Molecules → Organelles → Cells → Tissues → Organs → Organ Systems → Organisms

51. What does the diagram describe?

    A. the functioning of organs in the human body
    B. levels of organization in a living organism
    C. the process of evolution
    D. how inherited traits are expressed

Refer to the following information to answer Questions 52–54.

Immunity from a specific disease means that a person has the ability to resist that disease. Five types of immunity are listed below.

- **Inherited immunity**—Immunity that is inherited and may be permanent
- **Naturally acquired active immunity**—An often permanent immunity that occurs when a person's body naturally produces antibodies after being exposed to or infected by a disease
- **Naturally acquired passive immunity**—An immunity lasting for one year or less that occurs in a fetus or small infant because of antibodies passed to the offspring by the mother
- **Artificially acquired active immunity**—A long-term immunity that occurs when a person produces antibodies after being injected with a vaccine containing dead or weakened disease-causing pathogens
- **Artificially acquired passive immunity**—Usually a short-term immunity that occurs when antibodies produced in an animal are injected into a person

52. Studies show that children who were breastfed as babies are less likely to get certain diseases than children fed with baby formula. Research has shown that mother's milk provides antibodies that are not present in formula. What type of immunity is provided by mother's milk?

    A. naturally acquired active immunity
    B. naturally acquired passive immunity
    C. artificially acquired active immunity
    D. artificially acquired passive immunity

53. For protection against tetanus, an often-fatal disease caused by bacteria, doctors recommend that children should be given a tetanus shot at about 2 months of age. The antibodies present in a tetanus shot are produced by a horse. What type of immunity does a tetanus shot provide?

    A. naturally acquired active immunity
    B. naturally acquired passive immunity
    C. artificially acquired active immunity
    D. artificially acquired passive immunity

54. The natural concentration of antibodies in a person's bloodstream changes when the body is exposed at two different times to the same disease-causing organism. The primary immune response is often not sufficient to keep the person from getting the disease, but the secondary immune response may be sufficient. What type of immunity is this process?

    A. naturally acquired active immunity
    B. naturally acquired passive immunity
    C. artificially acquired active immunity
    D. artificially acquired passive immunity

Refer to the following information to answer Questions 55–57.

Photosynthesis is a process used by autotrophs to convert light energy, normally from the sun, into chemical energy that can later be used for the metabolic activities.

The general chemical equation for photosynthesis is:

$$6CO_2 + 6H_2O + \text{light energy} \rightarrow C_6H_{12}O_6 + 6O_2$$

carbon dioxide + water + light energy → sugar + oxygen

An experiment was conducted in which one bean plant was placed in a sunny window and the other was placed in a dark room. Both plants were supplied with equal amounts of water. After one day, one leaf from each plant was tested for the presence of sugar.

55. What is the most likely outcome of the test?

    A. The plant that was kept in the dark room will die.
    B. The leaf from the plant kept in the dark room will show the presence of sugar because there is enough water and carbon dioxide.
    C. Only the leaf from the plant in the window will show signs of sugar because sunlight is necessary for photosynthesis.
    D. Neither leaf will show signs of the presence of sugar, because one day is not enough time for photosynthesis to occur.

56. Which of the following is the most important step in this experiment?

    A. placing the bean plants in identical pots
    B. keeping both the plants in the dark for 24 hours so that previously produced sugars are not present
    C. using distilled water so that the minerals or other chemicals do not hinder the results
    D. ensuring that no worms are present in the soil

Write your answers in the blanks.

57. The end products of photosynthesis are ________________ and ________________.

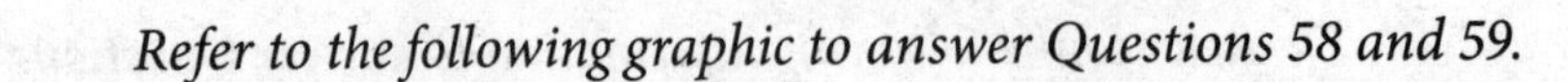

Refer to the following graphic to answer Questions 58 and 59.

The picture represents an ecosystem.

58. What is the primary source of energy in the ecosystem?

    A. a stream that starts in the mountains
    B. the sun
    C. the insects in the grass and trees
    D. the smaller animals such as rabbits and squirrels

59. The wolves in the ecosystem hunt for rabbits, which eat the plants present in the ecosystem. This is an example of

    A. cannibalism
    B. a food web
    C. a food chain
    D. photosynthesis

Use the following information to answer Question 60.

Bacteria can be classified by whether they need oxygen to live.

- **Aerobic** bacteria can survive only in places where oxygen is plentiful.
- **Anaerobic** bacteria are found only where oxygen is limited or entirely absent.
- **Facultative anaerobes** can survive with or without oxygen.

60. In which places would aerobic bacteria be able to survive?

    A. in fermenting wine
    B. in the upper atmosphere
    C. in infected human lungs
    D. inside a rotten piece of wood

Use the following information to answer Questions 61–63.

A plant scientist developed a new hybrid grass by crossing a desert grass with a shade grass. To determine the hybrid's ideal growing conditions, an experiment is performed in which the hybrid is grown in four different planters. Each planter is provided sunlight and water as shown in the table below. The average height of the grass in each planter is recorded weekly.

Planter	Watering Schedule	Amount of Sunlight
Planter A	3 times per week	8 hours or more per day
Planter B	3 times per week	4 hours or less per day
Planter C	once every 2 weeks	8 hours or more per day
Planter D	once every 2 weeks	4 hours or less per day

61. Which of the following conditions, if not met, would LEAST affect the results of the experiment?

    A. The depth of the soil should be identical in four planters.
    B. The planters should be weed-free.
    C. The same type of soil should be placed in all four planters.
    D. The four planters should be kept on the same table.

62. Which two planters should be compared to determine the effect of water on the growth of the hybrid grass when it receives the maximum amount of sunlight?

    A. A and B
    B. A and D
    C. C and A
    D. C and B

Write your answers in the blanks.

63. Identify the independent and dependent variables in this experiment.

    Independent: ____________________ and ____________________.

    Dependent: ____________________.

64. A dentist wants to prevent the possible spread of hepatitis C in her dental office. Which of the following steps is the LEAST practical precaution she can take?

    A. Wear disposable plastic gloves.
    B. Refuse to do dental work that causes gums to bleed.
    C. Disinfect all dental instruments before each use.
    D. Place all blood-soaked items in a plastic bag for safe disposal.

Use the following information to answer Questions 65–67.

The following diagram is known as the energy pyramid. It represents the graphical model of flow of energy in a community. Each level in the energy pyramid is known as a trophic level. Approximately 10 percent of the energy passes on from one trophic level to another.

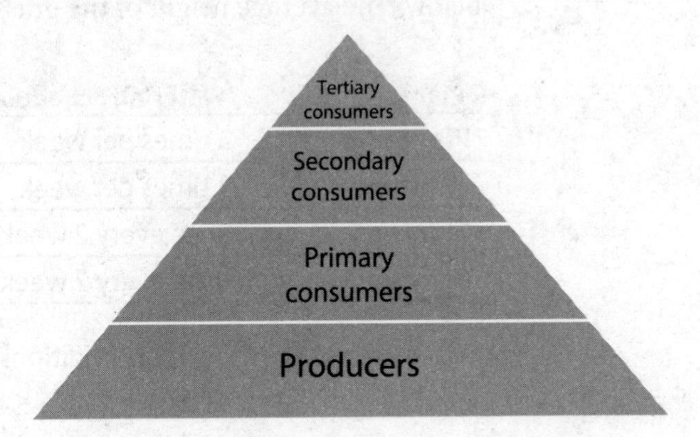

65. Which of the following information is NOT represented in the pyramid?

    A. The number of organisms decreases as you move up the pyramid.
    B. The energy flow in the ecosystem starts from the producers.
    C. If there is a sudden increase in the population of primary consumers, then that block will become the lowest in the pyramid.
    D. Energy flows from the bottom to the top of the pyramid.

66. Suppose the producers had 18,000 units of energy. Approximately how much energy would reach the level of tertiary consumers?

    A. 1.8 units
    B. 180 units
    C. 1,800 units
    D. 18 units

67. Which of the following is a significant factor of the energy transfer depicted by the energy pyramid?

    A. The energy pyramid is applicable to animals only.
    B. There is a large difference in the number of organisms in the lowest and highest trophic levels.
    C. The energy pyramid does not include all living organisms in the ecosystem.
    D. The energy pyramid holds true only during the daytime, i.e., when sunlight is present.

Refer to the following information to answer Questions 68 and 69.

A **food chain** is a sequence of energy transfer from one group of organisms to another in the ecosystem. Several interconnected overlapping food chains form a **food web**.

A typical food chain consists of the following:

- **Producers**—The base of the food chain is formed by autotrophic primary producer organisms. They produce their own food, usually through the process of photosynthesis.
- **Primary consumers**—These organisms do not produce food themselves but depend on primary producers for food.
- **Secondary consumers**—These animals depend on other animals, mainly herbivores, for food.
- **Tertiary consumers**—These animals take their food either directly or indirectly from all other trophic levels.

Each of the following two questions contains a blank marked Select . . . ▾. *Beneath it is a set of choices. Indicate the choice that is correct and belongs in the blank. (***Note:** *On the real GED test, the choices will appear as a "drop-down" menu. When you click on a choice, it will appear in the blank.)*

68. Carrion-eaters such as vultures and burying beetles are

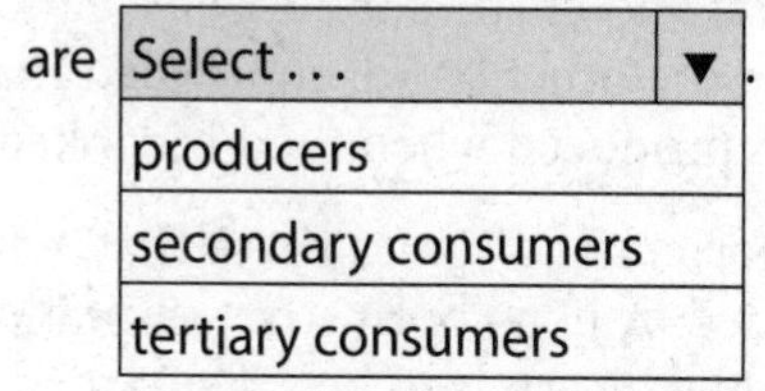

69. Grass is a

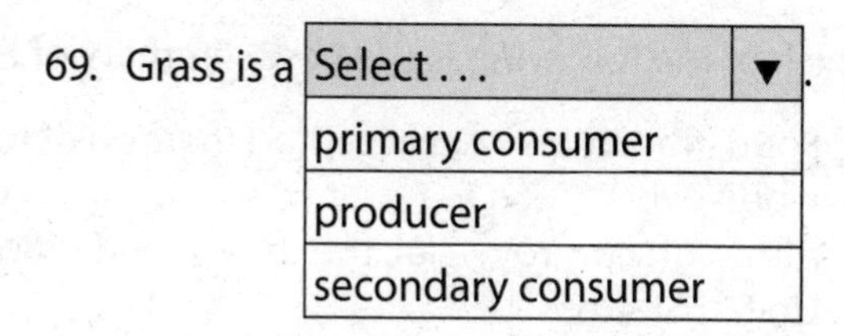

Questions 70–72 refer to the following passage.

A controversial health issue today is the use of irradiation to preserve food. In this process high-energy gamma rays (obtained from the radioactive decay of cobalt) or high-energy X-rays are used to kill mold; insects, such parasites as trichina worms in pork; and bacteria such as *Salmonella*.

Food irradiation has been used in dozens of countries, but it has only slowly come into use in the United States. The US Food and Drug Administration (FDA) first approved irradiation for harvested wheat and potatoes in the early 1960s. No other foods were approved until the 1980s, when spices, pork, fruits, and vegetables were added to the list. More recently, beef was added to the approved list in 1997. Many other foods are expected to be approved for irradiation in the coming years.

Although gamma rays are a form of nuclear radiation, irradiated food is not radioactive. It can be eaten immediately, or it can be vacuum-packed and stored safely for several years. Supporters claim that irradiation is a safe alternative to pesticides and more traditional preservatives.

However, critics point out that because irradiation can be used only on harvested crops, pesticides will still be needed to protect crops in the fields. What's more, they say that irradiation may kill odor-producing organisms that signal spoilage without killing the sources of food poisoning. Thus irradiation may destroy an important natural warning system.

Some critics also have claimed that irradiation may change the chemical makeup of food and create carcinogens (cancer-causing chemicals). Supporters, on the other hand, contended that an equal amount of carcinogens is produced when food is cooked or frozen.

Both supporters and critics agree that, with regard to the handling of the food—irradiated or not—both strict sanitation and cooking standards are essential if public health is to be safeguarded.

70. Which of the following is the best summary of the passage?

    A. Food irradiation is an accepted method of food preservation the world over.
    B. Although controversial, the use of food irradiation is increasing in the United States.
    C. Food irradiation does not result in deadly radioactive waste.
    D. Food irradiation can bring down the costs of dried foods.

71. When fruit is irradiated for a very long time, it turns squishy. What is one possible effect on fruit of long-term exposure to gamma rays?
    A. the breaking down of the fruit's chromosomes
    B. the breaking down of the fruit's cell walls
    C. the changing of the fruit's color
    D. the destruction of the fruit's cellular nuclei

72. Which of the following facts is LEAST important to FDA food scientists who are trying to determine what, if any, are the long-term dangers of food irradiation?
    A. So far, no adverse reactions to irradiation have been observed.
    B. Cancer associated with food irradiation may take 20 or more years to develop.
    C. The exact chemical changes caused by gamma radiation are impossible to determine.
    D. Because of the well-publicized dangers of radioactive waste, many people fear the food irradiation process.

Questions 73 and 74 refer to the following table.

The following table shows the risk factor of various birth-control methods when instructions are carefully followed. All methods are even less effective when not used properly.

Birth-Control Method	Risk Factor (approximate percentage of women who become pregnant each year while using this method)
Male condom	16%
Female condom	21%
Female diaphragm / cervical cap	18%
Spermicide	30%
IUD	4%
Birth-control pill	6%
Morning-after pill	25%
Sterilization (tubal ligation or vasectomy)	Almost 0%
Rhythm method	19%
Withdrawal method	24%
Abstinence	0%

73. Of the following birth-control methods listed in the table, which is the most effective?

    A. male condom
    B. spermicide
    C. birth-control pill
    D. sterilization

74. In which of the following publications or advertisements would you be LEAST likely to find this table?

    A. an advertisement for female condoms
    B. a community health-clinic brochure
    C. a book on different forms of birth control
    D. an advertisement for birth-control pills

75. Saccharin is a chemical substance that is about 500 times sweeter than ordinary sugar. Saccharin is used as a sugar substitute in many foods and beverages. In 1978 following the discovery that saccharin increases the incidence of bladder cancer in rats, the US Food and Drug Administration (FDA) began requiring that all products containing saccharin carry a warning label. Which comparative study described below could provide evidence that saccharin increases the risk of bladder cancer in humans?

    A. a study of the exercise habits of a group of people who are bladder cancer patients
    B. a study of the long-term health of a group of healthy people who do not eat saccharin
    C. a study of the long-term health of a group of healthy people who eat saccharin regularly
    D. a study of the present eating habits of a group of people who are bladder cancer patients

Use the following information to answer Questions 76–78.

The term **symbiosis** is used when two species live and associate closely together and depend on each other. There are three symbiotic relationships—commensalism, parasitism, and mutualism.

- **Commensalism** is a symbiotic relationship in which one species benefits from the relationship, while the other is not affected.
- **Parasitism** is a symbiotic relationship in which one species benefits from the relationship and the other is harmed by it.
- **Mutualism** is a symbiotic relationship in which both species benefit from the relationship.

Each of the following three questions contains a blank marked Select . . . ▼. *Beneath it is a set of choices. Indicate the choice that is correct and belongs in the blank. (**Note**: On the real GED test, the choices will appear as a "drop-down" menu. When you click on a choice, it will appear in the blank.)*

76. Tapeworms attach themselves to the intestines of animals such as cows and pigs, and feed on the partly digested food from the host's intestines, thereby depriving the host of nutrients. This relationship can be classified as Select . . . ▼.

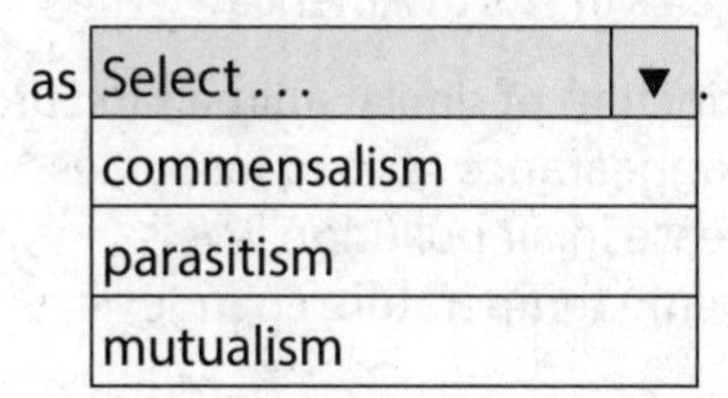

77. The remora fish forms a symbiotic relationship with sharks. The remora has special suckers attached to its fins that help it attach itself to the bodies of sharks, and it uses the shark for transportation as well as protection from predators. It also eats up the scraps of food that are left over when the shark eats its prey. This relationship can be classified as Select . . . ▼.

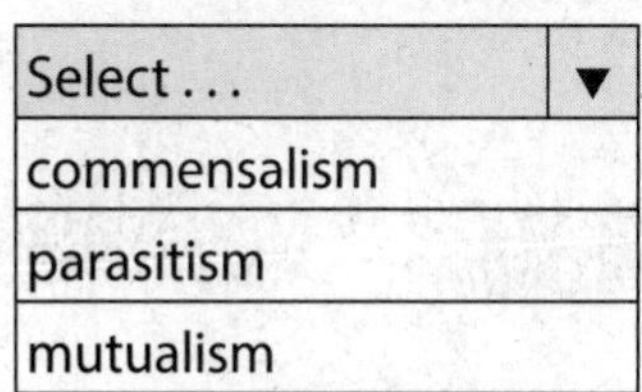

78. The honeybee sucks nectar from flowers, and in the process, pollen grains from the flower stick to the body of the bee. When the bee goes to suck nectar from another flower, the grains get transferred to that flower, thereby helping in the reproduction of flower-bearing plants. This relationship can be classified as Select . . . ▼.

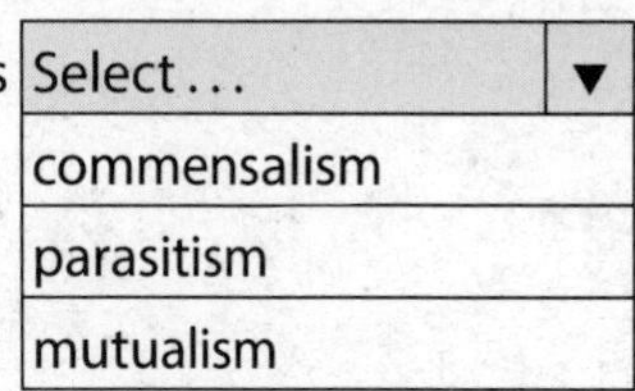

Use the following passage to answer Question 79.

The Central American acacia tree and the ant *Pseudomyrmex ferrugineus* live in a symbiotic relationship. The tree provides sweet nectar for the ants, and the ants protect the tree from weeds and animals. Investigations have revealed that the ants become addicted to the nectar and lose the ability to digest any other food! It was found that the ants are born with invertase, an enzyme that breaks down sugars. But invertase was not present in the adults. A single sip of the nectar causes irreversible damage to the invertase in the ants. Thus these ants can then survive only on the partially digested sugar of acacia nectar.

79. What is the most likely reason that the acacia nectar is partially digested by the ants?

    A. The ants do not like the taste of undigested sugar.
    B. The ants lack invertase and would be unable to fully digest the sugar.
    C. The ants cannot metabolize any other form of sugar.
    D. Ant reproduction is not possible without partially digested sugars.

80. Which of the following factors is LEAST likely to lead to the extinction of an animal species in North America?

    A. the extinction of similar animals in Europe
    B. the disappearance of food resources
    C. an increase in air pollution levels
    D. an extreme temperature change

Questions 81–84 refer to the following passage.

Medical researchers have discovered that smoking during pregnancy can be very harmful to a developing fetus. The fetus of a mother who smokes two packs of cigarettes each day loses 40 percent of the oxygen that would normally be present. This decrease in oxygen causes a baby born to a mother who smokes to weigh about 6 to 7 ounces lighter than a baby born to a nonsmoker.

The risk of having a stillborn child increases for a mother who smokes. Smoking one pack each day increases fetal death rate by 20 percent. Smoking two or more packs each day increases fetal death rate by more than 35 percent. Mothers who smoke are twice as likely to have miscarriages and are more likely to suffer other complications than are nonsmokers during pregnancy, including premature births.

Smoking at any time during pregnancy also increases the chance that a fetus will develop a malformed heart or other organ. Also, smoking doubles the chance for *abruptio placentae*—a premature separation of the placenta from the uterine wall, which may result in the death of the fetus.

The potential for harm to the child of a smoking mother does not end at birth. Young children of smokers are about 50 percent more likely to die of crib death and twice as likely to develop a lung illness or respiratory allergies as are children of nonsmokers.

81. On average, how much lighter is the newborn of a woman who smokes during pregnancy than is the newborn of a nonsmoker?

    A. 2 to 3 ounces
    B. 4 to 5 ounces
    C. 6 to 7 ounces
    D. 8 or more ounces

82. What is *abruptio placentae*?

    A. a complication occurring during the first 3 months of pregnancy
    B. a birth defect in the fetus
    C. a diminished supply of oxygen to the fetus due to smoking
    D. premature separation of the placenta from the uterine wall

83. Which of the following is NOT mentioned as a possible adverse effect of smoking while pregnant?

    A. miscarriage
    B. premature birth
    C. brain damage
    D. low birth weight

84. Babies born to mothers who smoke have a lower than average birth weight due to

    A. the presence of nicotine in the mother's blood
    B. an insufficient supply of oxygen to the fetus
    C. the increased risk of *abruptio placentae*
    D. the reduced health of the mother due to smoking

85. In eukaryotic organisms, such as plants and animals, most of the deoxyribonucleic acid (DNA) is located in a cell's

    A. nucleus
    B. vacuoles
    C. mitochondria
    D. ribosomes

The following question contains a blank marked Select . . . ▼. *Beneath it is a set of choices. Indicate the choice that is correct and belongs in the blank. (**Note**: On the real GED test, the choices will appear as a "drop-down" menu. When you click on a choice, it will appear in the blank.)*

86. Meiosis, unlike mitosis, produces daughter cells with

Select . . . ▼	chromosomes as the parent cell.
half as many	
the same number of	

Use the following information to answer Questions 87 and 88.

Animal cells are surrounded by a thin selectively permeable membrane known as the cell membrane, which forms a covering for the cell. Plant cells have an outer, rigid layer known as the cell wall. Inside the cell is a semi-liquid substance known as the cytoplasm. All metabolic activities of the cell occur here. Several organelles are also found here. Most cells also contain a membrane-bound nucleus. DNA, the genetic material that is duplicated before new cells form, is found within the nucleus.

Write your answer in the blank.

87. The ______________ gives shape to a plant cell.

88. Mitosis refers to the start of the process of cell division, in which the cell divides into two identical cells. Which part of the cell undergoes mitosis?

    A. nucleus
    B. cell membrane
    C. organelles
    D. mitochondrion

89. The human body is made up from several levels of organization. Select the correct sequence of the levels of organization.

    A. Cell < Organ < Tissue < Organ system
    B. Tissue < Cell < Organ system < Organ
    C. Cell < Tissue < Organ < Organ system
    D. Organ system < Organ < Tissue < Cell

Refer to the following information to answer Questions 90 and 91.

The cell cycle includes interphase, mitosis, and cytokinesis. Mitosis is a process of the cell cycle in which a cell's nucleus divides. The final result of mitosis and cytokinesis, which is when the cell divides, is two genetically identical daughter cells.

The four stages of mitosis are listed below (NOT in the order in which they occur).

- **Metaphase**—The chromosomes line up in a single file within the nucleus in the center of the cell.
- **Telophase**—The chromosomes begin to uncoil and a nuclear membrane forms around each set of chromosomes at opposite ends of the cell, forming two identical nuclei.
- **Anaphase**—The two sister chromatids of each chromosome separate. The now single-stranded chromosomes begin to move to opposite sides of the cell and the cell elongates.
- **Prophase**—The duplicated chromosomes condense into visible rods and the nuclear membrane breaks.

90. Based on the information provided, which is the correct sequence of the phases?

    A. Metaphase → Prophase → Telophase → Anaphase
    B. Anaphase → Metaphase → Prophase → Telophase
    C. Prophase → Metaphase → Anaphase → Telophase
    D. Anaphase → Prophase → Metaphase → Telophase

91. How many cells would be present after four complete mitosis cycles?

    A. 16
    B. 32
    C. 4
    D. 8

Use the following information to answer Questions 92 and 93.

Most bacteria divide by a process called binary fission. In binary fission a bacterium first produces a copy of its own DNA. The cell grows and splits into two parts, each part receiving a complete copy of its DNA.

92. Which of the following does NOT occur during binary fission?

    A. cell growth
    B. genetic diversity
    C. DNA replication
    D. increase in number of cells

93. On average, 1 cubic foot of air contains around 100 bacteria. About how many air-borne bacteria are present in a room that is 40 feet long, 30 feet wide, and 12 feet high?

    $$\text{volume} = \text{length} \times \text{width} \times \text{height}$$

    A. 14,400 bacteria
    B. 144,000 bacteria
    C. 1,440,000 bacteria
    D. 14,400,000 bacteria

Use the following information to answer Questions 94 and 95.

Meiosis is a special type of cell division that takes place only in the sex cells, that is, in the eggs and sperm for humans. This results in four haploid cells, containing half the number of chromosomes of body cells.

94. Human body cells consist of 46 chromosomes. After a successful cycle of meiosis, how many chromosomes would the daughter cells contain?

    A. 13 chromosomes
    B. 23 chromosomes
    C. 46 chromosomes
    D. 92 chromosomes

95. Which of the following is true?

    A. The number of daughter cells in mitosis is one-fourth the number of daughter cells formed in meiosis.
    B. The number of daughter cells in mitosis is double the number of daughter cells formed in meiosis.
    C. The number of daughter cells in mitosis is half the number of daughter cells formed in meiosis.
    D. The number of daughter cells in mitosis is triple the number of daughter cells formed in meiosis.

Questions 96–98 are based upon the following information.

A plant captures light energy and uses it to produce glucose. The reaction for this process is:

$$\text{sunlight} + 6CO_2 + 6H_2O \rightarrow 6O_2 + C_6H_{12}O_6$$

The glucose produced can then be stored in the plant as starch. A rabbit then consumes the plant and uses the starch from the plant as energy. After eating the plant, the rabbit is preyed upon by an eagle. The eagle takes the rabbit's carcass to its nest to feed both itself and its offspring.

96. The reaction shown in the passage is best described as

A. cellular respiration
B. a food chain
C. photosynthesis
D. ATP synthesis

97. A reaction that is most likely to occur within the rabbit, the eagle, and the eagle's offspring is

A. $\text{sunlight} + 6CO_2 + 6H_2O \rightarrow 6O_2$ and $C_6H_{12}O_6$
B. $6O_2$ and $C_6H_{12}O_6 \rightarrow 6CO_2 + 6H_2O + \text{heat energy}$
C. $4\,{}^{1}_{1}H \rightarrow {}^{4}_{2}He + 2\,{}^{0}_{1}e + \text{energy}$
D. ${}^{1}_{0}n + {}^{235}_{92}U \rightarrow {}^{141}_{56}Ba + X + {}^{31}_{0}n + \text{energy}$

98. The oxygen produced by the reaction in the passage can then be utilized by

A. an anaerobic organism
B. an aerobic organism
C. both an anaerobic organism and an aerobic organism
D. neither an anaerobic nor an aerobic organism

Use the following information to answer Question 99.

DNA, deoxyribonucleic acid, is the basic life chemical. DNA is a double-stranded structure made of four types of nitrogen bases, namely, adenine (A), thymine (T), guanine (G), and cytosine (C). The base sequence of a single strand corresponds to the sequence of the other strand. Adenine (A) always pairs with thymine (T), and cytosine (C) always pairs with guanine (G).

99. If the base sequence of one strand is A-T-C-G-G-A-C-T. What is the base sequence of the opposite strand?

A. G-C-A-T-T-G-C-T
B. T-A-G-C-C-T-G-A
C. A-T-C-G-T-C-A-G
D. C-G-A-T-T-C-A-G

Use the following information to answer Questions 100–102.

The actual allele pairs of an organism are called the genotype. The outward expression of an allele pair of an organism is called the phenotype.

Pea plants have spherical seeds (S) and dented seeds (s). A heterozygous pea plant (Ss) is crossed with a homozygous recessive pea plant (ss).

100. Fill the Punnett square based on the information in the passage.

	s	s
S		
s		

101. What is the phenotype ratio of this cross?
    A. 1 : 2 : 1
    B. 3 : 1
    C. 2 : 2
    D. 2 : 1 : 1

102. Which of the following is NOT a conclusion that you can draw from the illustration?
    A. The gene inherited from the male sperm is dominant over the gene inherited from the female egg.
    B. The number of genes in a sperm cell or egg cell is half the number of genes in the regular parent cells.
    C. Traits are passed down from one generation to the next.
    D. A child inherits genes from its parents, while a child's parents inherited genes from the child's grandparents.

103. Mutation is any change in the structure of the gene. The result of a mutation may benefit an organism or may have detrimental effects. Which is LEAST likely to be a reason for mutation?
    A. ultraviolet rays
    B. nuclear radioactivity
    C. chemical mutagens
    D. physical exercise

Question 104 refers to the following passage.

Alzheimer's disease is a progressive brain disorder that starts with a gradual memory loss. As the disease progresses, the patient loses language, perceptual, and motor skills. Eventually Alzheimer's patients lose the ability to care for themselves.

No one knows what causes Alzheimer's disease. It is known that Alzheimer's is most common in older patients and family history plays a role. Also, the disease involves distinctive formations in brain cells. These formations cause brain cells to shrink and die, leaving gaps in the brain's messaging network. The inability of the brain to function in a normal way gives rise to the symptoms of Alzheimer's.

104. Which of the following results of Alzheimer's disease can NOT be inferred from the passage?

A. the uncertainty about the patient's future
B. the likely death of the patient within 10 years
C. the cost of expensive medicines for the patient
D. the emotional difficulties that family members experience

Questions 105 and 106 refer to the following passage.

At the beginning of the twenty-first century, there is new hope for cancer patients. New drugs, such as Imatinib, fight cancer in a novel way. Taken as a pill, Imatinib targets specific leukemia cells for destruction and leaves healthy cells alone. In early testing, Imatinib is showing remarkable promise in stopping the progression of this type of cancer.

Before the invention of targeted cancer-cell fighters, standard treatments for cancer were surgery, radiation, and chemotherapy. Surgery almost always leaves behind cancer cells, and radiation and chemotherapy can kill more healthy cells than cancer cells. What's more, these traditional cancer treatments are not effective against many types of cancer.

The hope now is to develop other cancer target-seeking drugs that will be able to seek out and destroy various types of cancer cells before the cancer harms the patient.

105. Which phrase does NOT describe Imatinib?

   A. part of a new strategy in fighting cancer
   B. can be taken in pill form
   C. leaves healthy cells alone
   D. a sure cure for many cancers

106. What may be one advantage of radiation?

   A. There are medicines available for the side effects of radiation treatment.
   B. Radiation treatment is covered by insurance.
   C. The long-terms effects are known.
   D. There is a statistical probability for a total cure.

107. In any ecosystem organisms compete for limited resources. Competition is most severe between members of the same species. However, it also occurs between organisms that are similar but not identical. Between which pair of animals is competition LEAST likely to occur?

   A. a whale and a dolphin
   B. a squirrel and a chipmunk
   C. a deer and a blackbird
   D. a sparrow and a finch

Use the following information to answer Questions 108–113.

The orangutan is the most intelligent of all land animals. Its name comes from the Malay words *orang* and *hutan*, which means "man of the forest." The Indonesian islands are the only places where these "forest men" live.

The adult male orangutans of Borneo roam freely and do not stay close to any family unit. A few miles away on Sumatra, however, an adult male does stay close to his mate and offspring. Differences in male orangutan behavior are most likely the result of learned or possibly inherited behavior patterns that best ensure the survival of orangutans in each location.

On the island of Borneo, male orangutans are not needed to help ensure the health and safety of the females and their young ones. The older, dominant males mate with any fertile receptive female. On Borneo, if a large male were to stay near his mate and offspring, he likely would have to compete for food and actually decrease his offspring's chance of survival.

Sumatra, however, contains leopards that prey on female and infant orangutans and siamangs, tree-dwelling apes that compete with orangutans for food. For both these reasons, the survival chance of the female and infant increases if the males are close.

108. With evidence from the passage, describe the behavioral difference of the orangutans of Sumatra and Borneo and the reasons for this difference.

Write your response on the lines.

109. On Sumatra, which animal is known to hurt orangutans?

A. siamangs
B. snakes
C. lions
D. leopards

110. What threat to the health of orangutans does the passage imply may exist on Borneo?

 A. leopards
 B. infertility
 C. starvation
 D. overpopulation

111. What general biological principle is exemplified by the male orangutans on Borneo and Sumatra showing different social behaviors?

 A. speciation
 B. adaptation
 C. reaction
 D. camouflage

112. What do the parenting behaviors of the orangutans on Borneo and Sumatra have in common?

 A. Both protect orangutans from human hunters.
 B. Both establish a social group of orangutans.
 C. Both produce the greatest number of surviving offspring.
 D. Both maintain closely bonded family units.

113. Which scientist's work would best help in the understanding of orangutan behavior?

 A. Archimedes
 B. Charles Darwin
 C. Albert Einstein
 D. Aristotle

114. As a general rule, animals living in the wild have a shorter life expectancy than animals raised in captivity. What is the most likely reason for this difference?

 A. Animals in the wild are more likely to be killed by predators.
 B. Animals in the wild age quickly.
 C. Animals in the wild eat non-nutritional food.
 D. Animals in the wild are subject to more stress.

Use the following information to answer Questions 115–119.

In a learning theory experiment, a group of octopuses was trained to attack a red ball. Another group was trained to attack a white ball. Meanwhile, two groups of untrained octopuses watched. One group watched the red-ball tank; the other watched the white-ball tank. When both red and white balls were placed in the tanks with the untrained octopuses, each octopus attacked only the color ball that was attacked in the group it had watched. Although octopuses have the largest brain amongst invertebrates, this result was surprising to many scientists. This social-learning skill had been considered beyond the ability of an octopus.

The parts of the learning theory experiment with the octopuses can be classified into these five categories:

- **Experiment**—A procedure used to investigate a problem
- **Finding**—An experimental result or a conclusion reached as a part of the investigation
- **Hypothesis**—A reasonable, but not proved, explanation of an observed fact
- **Prediction**—An opinion about something that may occur in the future
- **Nonessential fact**—A fact that does not directly help the researcher understand the problem being investigated

For Questions 115–119, write your answers in the blanks.

115. Octopuses that observed a red ball being attacked chose to attack a red ball themselves. This statement is best classified as ________________.

116. To see if octopuses can learn by observation, untrained octopuses were allowed to watch trained octopuses attack colored balls. This statement is best classified as ________________.

117. The learning ability of an octopus most likely arose from its need to learn from other octopuses at an early age, because the parents of an octopus die when it is hatched. This statement is best classified as ________________.

118. An octopus has a soft, oval body and eight arms and lives mostly at the bottom of the ocean. This statement is best classified as ________________.

119. With further research, scientists will find that octopuses have a wide range of learning abilities not yet discovered. This statement is best classified as ________________.

120. Nebraska's Sand Hills are home to a deer mouse that is one of the quickest evolving animals. The deer mouse is normally dark brown, which is a good color for mice living in the woods and surrounding areas since it allows them to hide better and avoid predators. The deer mouse that lives in the Sand Hills, however, has evolved into a much lighter, sandlike color. Without this change, the deer mouse would be easily spotted by predators against the area's light terrain. This is an example of

A. natural selection
B. extinction
C. artificial selection
D. invasion

121. Trout are known to survive best in cool water shaded by overhanging branches. Cool water contains more dissolved oxygen than warm water. Trout may die if they are constantly warmed by sunlight. How would global warming and deforestation affect trout population in a lake?

Write your response on the lines.

____

____

____

____

____

____

____

____

____

____

122. Seen from the top, trout are dark in color with a speckled appearance, similar to the bottom of a river or lake. Seen from the bottom, trout are lighter in color, similar to the water surface as seen by an underwater swimmer. What is this characteristic called?

A. behavior
B. camouflage
C. stimulus–response
D. hybridization

Use the following information to answer Questions 123–124.

The graph represents a predator–prey relationship. The solid line represents the relative number of cheetahs in an ecosystem, and the dashed line represents the relative number of gazelles present in the same ecosystem. The *x*-axis shows the number of years that have passed.

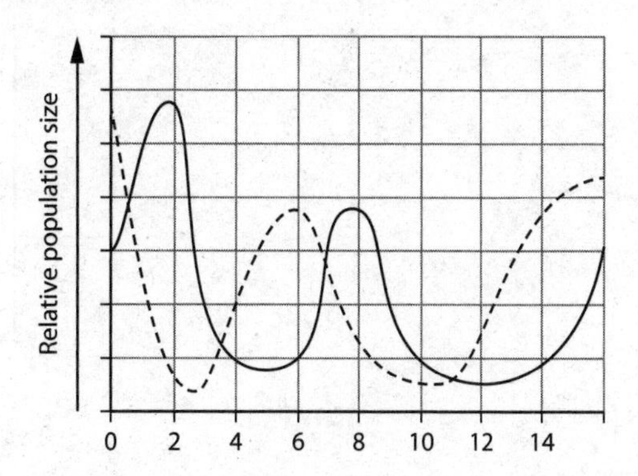

123. What is the most probable reason for the increase in the predator population from year 6 to year 8?

   A. an increase in the prey population from year 3 to year 6
   B. a predator population equal in size to the prey population from year 5 to year 6
   C. a decrease in the prey population from year 1 to year 2
   D. the extinction of the prey population in year 3

124. Place a dot on the graph at a point when predator and prey populations were at equilibrium.

125. A thermostat in your home controls the heating and cooling. For example, during the winter you might set the thermostat to 68°F to keep the temperature of the home constant. If the temperature drops below 68°F, a signal is sent to the furnace telling the furnace to produce heat until the inside temperature of the home is 68°F again. Similarly, when humans eat, blood sugar levels increase. This sends a signal to the brain that tells the body to release insulin into the blood in order to metabolize the added sugar. This feedback mechanism helps maintain

   A. only the person's body temperature
   B. a food web within the body
   C. a proper biome for humans to live in
   D. homeostasis in the body

Use the following diagram that shows the evolutionary relationships of some organisms to answer Questions 126–128.

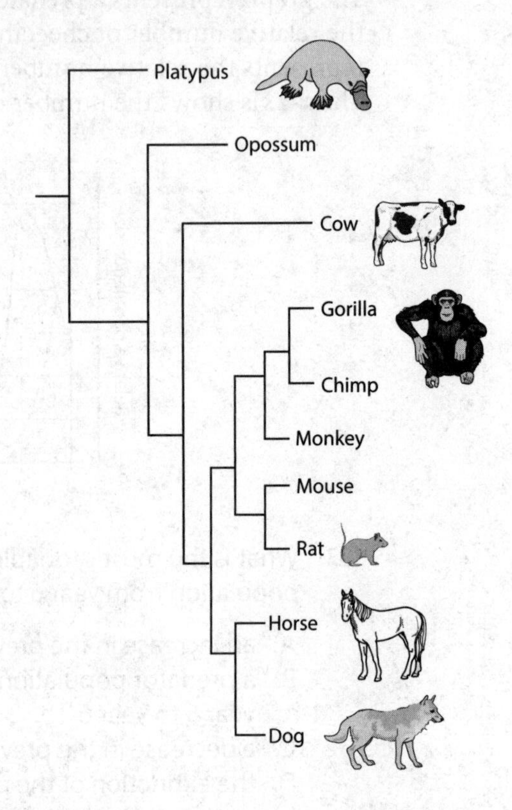

126. Which two organisms would most likely synthesize the most similar enzymes?

  A. monkey and mouse
  B. cow and horse
  C. chimp and rat
  D. horse and dog

127. Which of the following organisms are most closely related?

  A. platypus and opossum
  B. gorilla and monkey
  C. gorilla and chimp
  D. rat and dog

128. Place an X on the spot on the diagram that shows the most recent common ancestor of the cow and the horse.

129. Select the choice that is a type of external chemical barrier provided by the body's innate immune system.

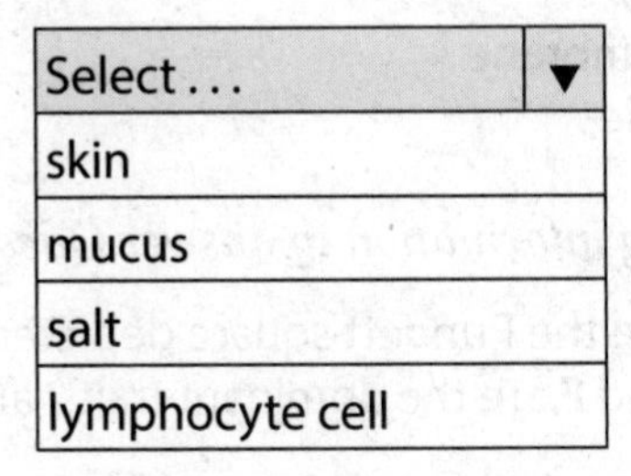

130. Which of the following describes a way in which the neuromuscular and skeletal systems work together?

   A. Muscles in the lungs contract and relax to inhale and exhale.
   B. Bones synthesize blood and immune cells and store minerals.
   C. Bones are connected to other bones via ligaments and to muscle fibers via tendons.
   D. Bones elongate and ossify as adolescents grow.

Use the following information to answer Questions 131–132.

131. Identify the cell structure marked in the following image.

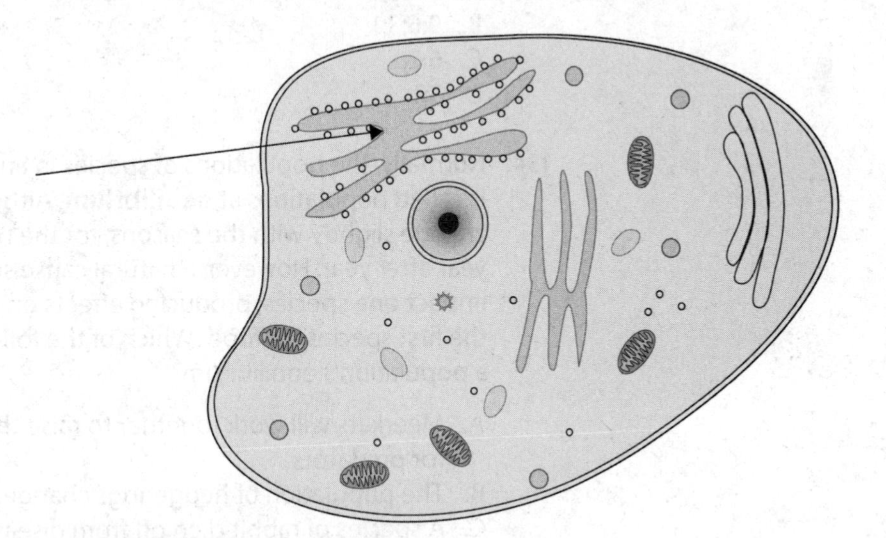

   A. smooth endoplasmic reticulum
   B. Golgi apparatus
   C. mitochondria
   D. rough endoplasmic reticulum

132. Which part of the cell controls what enters and exits the cell?

A. lysosomes
B. nucleus
C. cell membrane
D. centriole

Use the following information to answer Question 133.

A student created the Punnett square depicted here to predict the phenotypes of offspring. **R** and **F** are the dominant traits, and **r** and **f** are the recessive traits.

	RF	**Rf**	**rF**	**rf**
RF	RRFF	RRFf	RrFF	RrFf
Rf	RRFf	RRff	RrFf	Rrff
rF	RrFF	RrFf	rrFF	rrFf
rf	RrFf	Rrff	rrFf	rrff

133. What is the phenotypic ratio in this dihybrid cross?

A. 9:3:3:1
B. 9:6:3:1
C. 6:2:1
D. 3:1

134. Normally, the populations of species in an ecosystem remain stable. This is called populations at **equilibrium**. Although the populations might change slightly with the seasons, for the most part they stay the same year after year. However, a natural cause such as disease can suddenly impact one species, producing effects on other species that depend on the first species for food. Which of the following describes a disruption of a population's equilibrium?

A. Meerkats will work together to raise their young and keep an eye out for predators.
B. The population of hedgehogs changes slightly with the seasons.
C. A species of rabbit died off from disease, so the local wolves and hawks find it harder to find food, causing their populations to decline.
D. As the prey population increases, the predator population also increases.

135. The peppered moth is found in two varieties—light and dark. The moths live in trees, but in a certain area where they live, industrial development has caused the trees to become covered with soot. In this area, the number of dark moths is now much greater than the number of light moths. Which term describes this phenomenon?

A. natural selection
B. artificial selection
C. extinction
D. speciation

The following question contains a blank marked Select . . . ▼ *Beneath it is a set of choices. Indicate the choice that is correct and belongs in the blank. (**Note**: On the real GED® test, the choices will appear as a "drop-down" menu. When you click on a choice, it will appear in the blank.)*

136. The DNA nucleotide sequences of different organisms can be compared by using

Select . . . ▼
a microscope.
gel electrophoresis.
a family tree.

Use the following information to answer Questions 137–138.

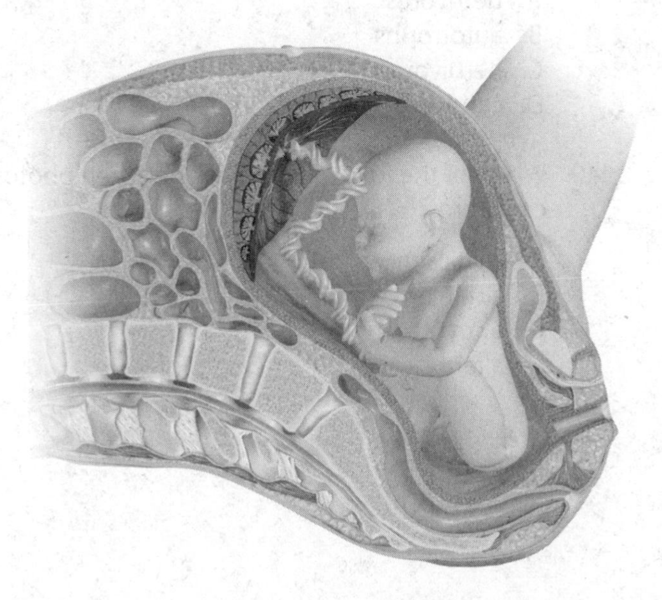

137. On the diagram above, draw a line pointing to the placenta. (***Note***: *On the real GED® test, you will simply click on a part of the diagram.*)

138. Which of the following is a function of the umbilical cord?

  A. carrying oxygenated blood to the fetus
  B. delivering milk to the fetus
  C. allowing for fetal respiration
  D. connecting the fetus to the mother

Use the following information to answer Questions 139–140.

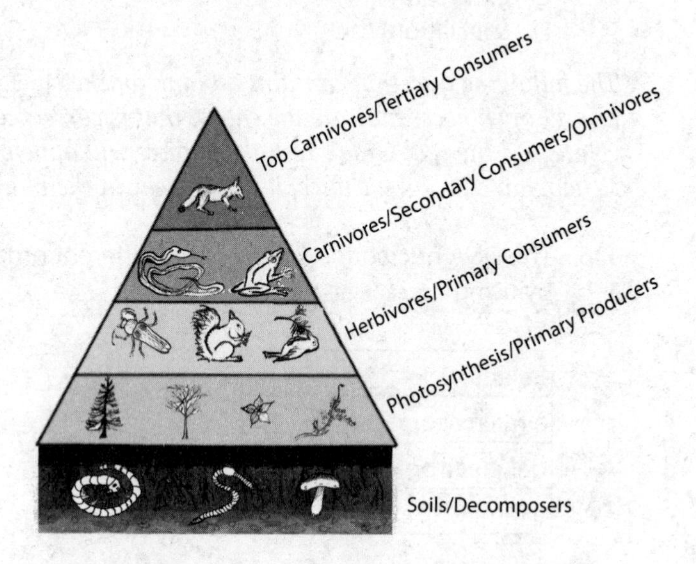

139. Which of the following accounts for the most energy in the ecological pyramid?

  A. detrivores
  B. autotrophs
  C. herbivores
  D. carnivores

140. Which of the following is a link between photosynthetic organisms and carnivores?

  A. a chicken
  B. a human
  C. a coyote
  D. a rabbit

CHAPTER 2

Physical Science

Topics from this chapter constitute about 40 percent of the GED Science test. This chapter covers three main topics. The first is conservation, transformation, and flow of energy. This topic includes sources, types, and transfer of energy in the world around us. The second topic is work, motion, and forces. Basic concepts of speed, velocity, Newton's laws of motion, scientific meanings of work and power, gravitational forces, and simple machines are explored. The third topic is chemical properties and reactions, covering information about atoms and molecules, their interactions and transformations, and related energy. Each of these science topics involves fundamental principles that govern the universe and its existence and interactions on Earth. This chapter is based on these topics and will help you with the physical science portion of the GED Science test.

Directions: Answer the following questions. For multiple-choice questions, choose the best answer. For other questions, follow the directions preceding the question. Answers begin on page 169.

Questions 1–3 refer to the following passage.

Heat is the transfer of thermal energy from one substance to another or within a substance. Thermal energy includes both potential and kinetic energy of the particles that make up the substance. A warmer substance becomes cooler and a cooler substance becomes warmer when heat flows from one to the other. This flow of heat is always from the substance with a higher temperature to the substance with a lower temperature.

1. Which of the following statements is correct?
   A. Heat flows from a substance of lower internal energy to a substance of higher internal energy.
   B. Heat flows from a liquid to a solid.
   C. Heat flows from a substance at 100°C to a substance at 50°C.
   D. Heat flows from a substance of greater mass to the substance of smaller mass.

2. Study the diagram.

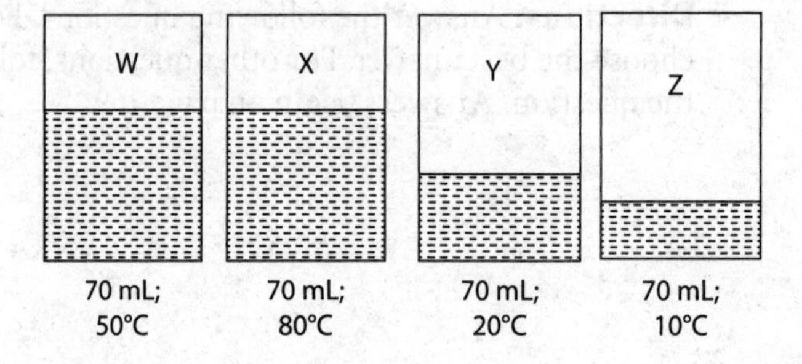

   Which one of the containers holds the water with the highest average kinetic energy?

   A. W
   B. X
   C. Y
   D. Z

*The following question contains a blank marked "Select... ▼." Beneath it is a set of choices. Indicate the choice that is correct and belongs in the blank. (**Note**: On the real GED test, the choices will appear as a "drop-down" menu. When you click on a choice, it will appear in the blank.)*

3. An iron ball at 40°C is dropped in a mug containing water at 40°C.

   Heat will Select... ▼.

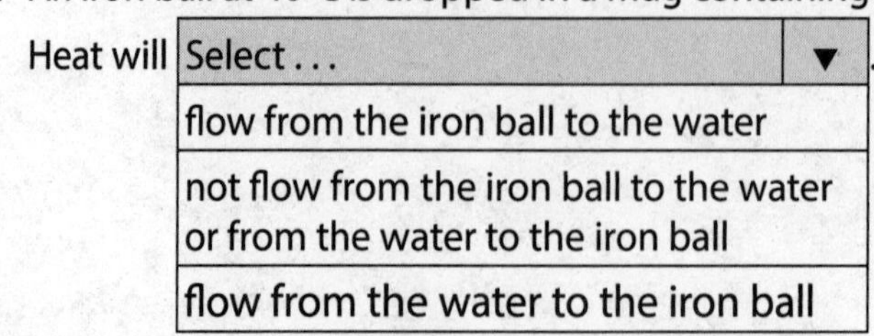

Questions 4–8 refer to the following passage.

Temperature is a measure of the average kinetic energy of the particles that make up a substance. Because it is not possible to measure the kinetic energy of individual particles of a substance and take the average, a thermometer is used to measure temperature. The following diagram shows the comparison between the various scales commonly used to measure temperature.

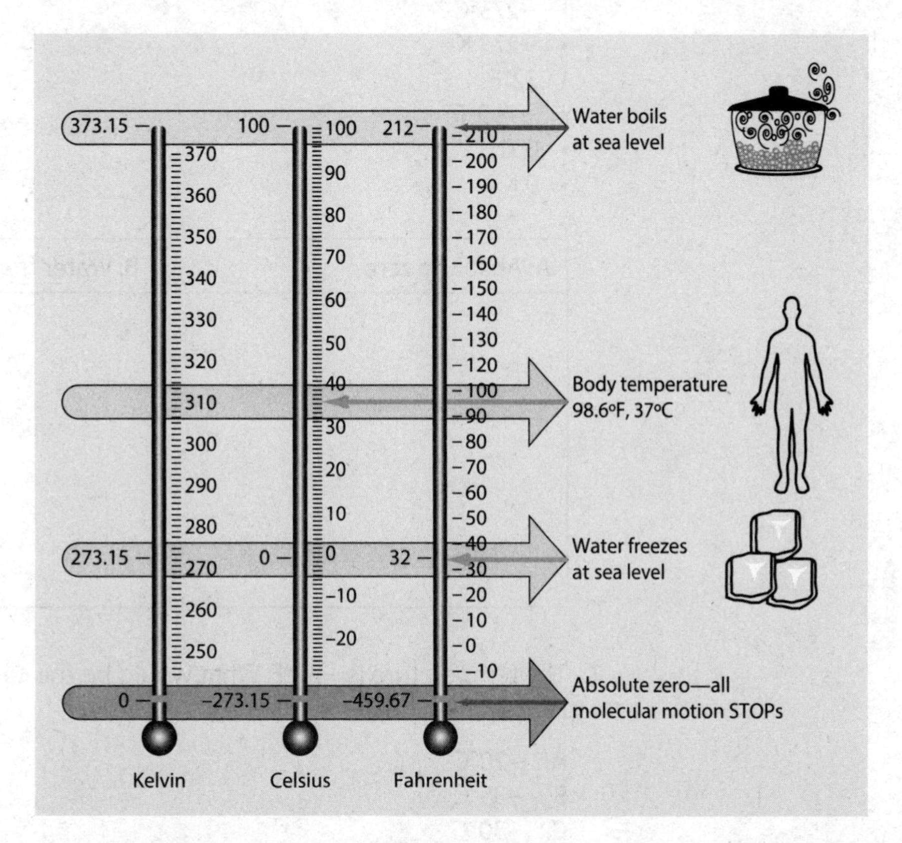

Scientists worldwide use the Celsius scale. It is based on the properties of water, and the unit for it is degrees Celsius, written as °C. On the Celsius scale, water freezes at 0°C and boils at 100°C. Kelvin is the SI unit of temperature, written as K. Absolute zero, 0 K, which is equivalent to −273°C, is the lowest temperature possible for any substance. The Fahrenheit scale is commonly used in the United States for weather reports and to measure temperatures indoors. The unit is degrees Fahrenheit, written as °F.

To convert between Fahrenheit and Celsius scales, any temperature y°F is converted to the temperature x°C with the following equation:

$$y°\mathrm{F} = \frac{9}{5}x°\mathrm{C} + 32$$

4. Which temperature has the same value on the Celsius and Fahrenheit scales?

   A. +40°
   B. −40°
   C. +100°
   D. −50°

5. At what temperature on the Fahrenheit scale does water boil?

   A. +100°
   B. +150°
   C. +212°
   D. +273°

6. Place each temperature in the correct box.

   - −273°C
   - +273 K
   - 32°F
   - −459°F
   - 0°C
   - 0 K

A: Absolute zero	B: Water freezes

7. The temperature is −22°F. What would be the corresponding temperature in the Celsius scale?

   A. −20°C
   B. −35°C
   C. −30°C
   D. −12°C

8. Which of the following temperature scales does NOT have negative numbers?

   A. Celsius
   B. Kelvin
   C. all scales
   D. Fahrenheit

Questions 9–11 refer to the following passage.

The specific heat (C) refers to the amount of heat required to cause a unit of mass (gram or kilogram) to change its temperature by 1°C. The standard metric unit is joules/kilogram/Kelvin (J/kg/K).

$$C = \Delta Q/(m \times \Delta T)$$

where C = specific heat, in J/(kg·K); ΔQ = heat energy required for the temperature change, in joules; ΔT = change in temperature (i.e., $T_{final} - T_{initial}$); and m = mass of the object, in kilograms.

Different substances have different specific heat. The table below shows some examples.

Substance	Specific Heat (J/kg·K)
Air	1,010
Aluminium	902
Copper	385
Gold	129
Iron	450
Mercury	140
Lead	128
Ice	2,030
Water	4,186
Salt	864

Due to its relatively high heat capacity, water plays a very important role in temperature regulation.

9. How much heat is required to raise the temperature of a 225-g lead ball from 15°C to 25°C?

   A. 217 J
   B. 725 J
   C. 288 J
   D. 145 J

10. Students conducted the following experiment:

Experiment

- Amount of heat Q was applied to 1 kg of aluminum.
- Amount of heat Q was applied to 1 kg of copper.

The heat was applied to the two metals under identical experimental conditions.

Observation

The students observed that the rise in temperature of the copper was much greater than the rise in temperature of the aluminum.

Which of the following results can be inferred from this experiment?

A. The rise in temperature of a substance is equal to its specific heat.
B. The rise in temperature of a substance is directly proportional to its specific heat.
C. The rise in temperature of a substance has no relation to its specific heat.
D. The rise in temperature of a substance is inversely proportional to its specific heat.

11. Explain why Lake Michigan remains quite cold even in early July despite the outside air temperatures being near or above 90°F (32°C).

Write your response on the lines.

Questions 12 and 13 refer to the following graph.

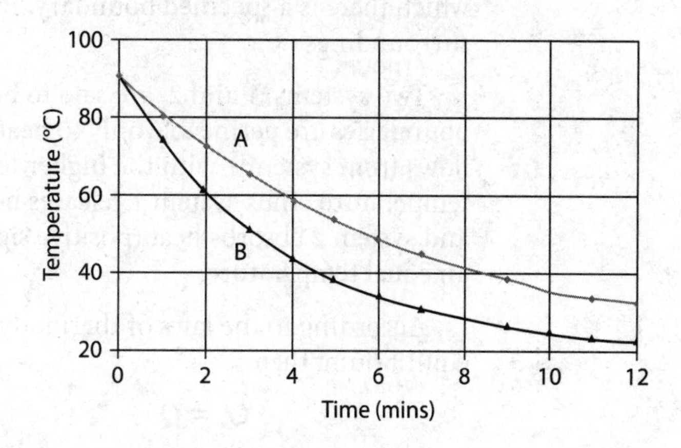

12. From the cooling curves, find the decrease in temperature of liquid A in the first six minutes.

    A. 40°C
    B. 50°C
    C. 55°C
    D. 35°C

*The following question contains a blank marked "Select . . . ▼." Beneath it is a set of choices. Indicate the choice that is correct and belongs in the blank. (**Note**: On the real GED test, the choices will appear as a "drop-down" menu. When you click on a choice, it will appear in the blank.)*

13. From the cooling curves, it is observed that Select . . . ▼ has the lower specific heat.

Select . . . ▼
liquid A
liquid B

Questions 14 and 15 refer to the following passage.

A thermodynamic system is a quantity of matter of fixed identity, around which there is a specified boundary. Everything outside the boundary is the surroundings.

Two systems, 1 and 2, are said to be in thermal equilibrium if their boundaries are permeable only to heat and do not change over time. Heat flows from system 1 with the higher temperature to system 2 with the lower temperature. Thus system 1 releases heat (negative sign for heat released) and system 2 absorbs heat (positive sign for heat absorbed) until they attain an equal temperature.

According to the laws of thermodynamics, if two systems are in thermal equilibrium, then

$$Q_1 = Q_2$$

$$\Rightarrow \quad m_1 \times C_1 \times (T_{final} - T_1) = m_2 \times C_2 \times (T_{final} - T_2)$$

where

for system 1, Q_1 = heat released from system 1; m_1 = mass; C_1 = specific heat; temperature = T_1;

for system 2, Q_2 = heat absorbed by system 2; m_2 = mass; C_2 = specific heat; temperature = T_2;

T_{final} = final temperature of the final system when systems 1 and 2 are in contact.

14. A 12.9-g sample of an unknown metal at 26.5°C is placed in a Styrofoam cup containing 50 g of water at 88.6°C. The water cools down, and the metal warms up until thermal equilibrium is achieved at 87.1°C. Assuming all the heat released by the water is absorbed by the metal and that the cup is perfectly insulated, determine the specific heat of the unknown metal. The specific heat of water is 4,186 J/(kg·K).

    A. 1401.03 J/(kg·K)
    B. 401.03 J/(kg·K)
    C. 4010.3 J/(kg·K)
    D. 40,103 J/(kg·K)

Write your answer in the blank.

15. When two bodies are in thermal equilibrium, their ________________ are the same.

Questions 16 and 17 refer to the following passage.

A change in the energy of a system occurs due to heat added to the system and work done by the system. This can be explained by the first law of thermodynamics, which states that in a thermodynamic process involving a closed system, the change in the internal energy is equal to the difference between the *heat* transferred to the system and the work done by it. Therefore, taking ΔU as a change in internal energy, you can write

$$\Delta U = Q - W$$

where Q = quantity of heat supplied to the system by its surroundings and W = work done by the system on its surroundings. Q is positive if heat is added to the system and negative if heat is removed; W is positive if work is done by the system and negative if work is done on the system.

16. If a system has 60 J of heat added to it, resulting in a change of internal energy of 40 J, what is the work done by the system?

    A. 50 J
    B. 20 J
    C. 30 J
    D. 60 J

17. An athlete doing push-ups performs 650 kJ of work and loses 425 kJ of heat. What is the change in the internal energy of the athlete?

    A. −225 kJ
    B. −1,075 kJ
    C. 1,075 kJ
    D. 225 kJ

Questions 18–21 refer to the following information.

A good example of a system that can do work is the gas confined by a piston in a cylinder, as shown in the diagram below.

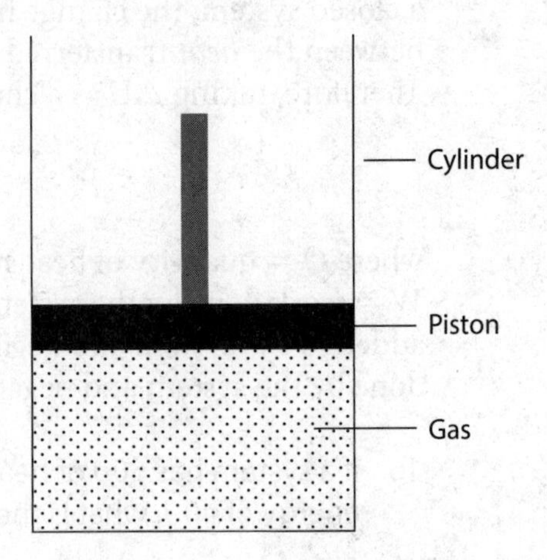

If the gas is heated, it will expand and push the piston up, thereby doing work on the piston. Alternatively, if the piston is pushed down, the piston does work on the gas and the gas does negative work on the piston. The work done is the measure of the pressure applied multiplied by the change in volume, that is,

$$W \text{ (work)} = P\Delta V$$

where W = work, P = pressure, and ΔV = change in volume.

If there is no change in the volume of the gas, that is, if $\Delta V = 0$, then no work is done.

A heat engine is any device that uses heat to perform work. There are three essential features of a heat engine:

- Heat is supplied to the engine at a high temperature from a hot reservoir.
- Part of the input heat is used to perform work.
- The remainder of the input heat is released into a cold reservoir, which is at a lower temperature than the hot reservoir.

In the diagram, if a block is placed on the piston in the gas-filled cylinder, heat can be used to do work in lifting the block. Any heat left over after work is done is released into the low temperature reservoir.

The percent efficiency, %*e*, of the heat engine is equal to the ratio of the work done to the amount of input heat:

$$\%e = \frac{\text{Work}}{Q_{\text{Hot}}} \times 100$$

18. Consider the following pressure vs. volume graph showing a complete thermodynamic cyclic process for a gas, from A → B, B → C, and C → A. If the work done going from C→ A is zero, then match A, B, and C with points 1, 2, and 3 on the graph.

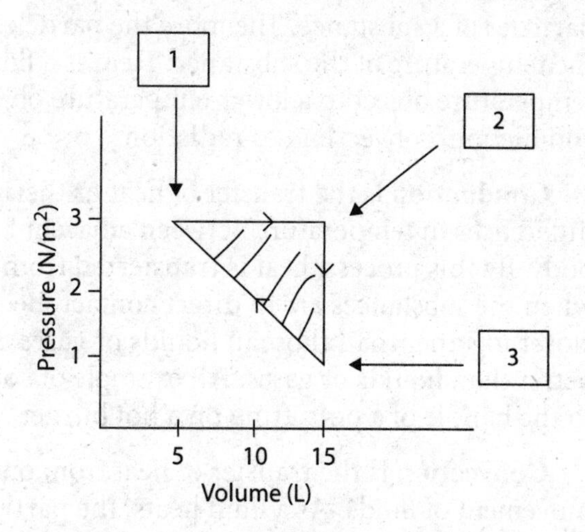

    A. Point 1 is B, point 2 is C, and point 3 is A.
    B. Point 1 is C, point 2 is B, and point 3 is A.
    C. Point 1 is A, point 2 is B, and point 3 is C.
    D. Point 1 is B, point 2 is A, and point 3 is C.

19. A gas in a container is at a pressure of 151,500 Pa and a volume of 4 L. What is the work done by the gas if it expands at constant pressure to twice its initial volume?
    A. 606 J
    B. 60,600 J
    C. 60.6000 J
    D. 606,000 J

20. A system has 60 J of heat added to it, resulting in 15 J of work being done by the system and exhausting the remaining 45 J of heat. What is the efficiency of this process?
    A. 100%
    B. 60%
    C. 45%
    D. 25%

21. When a heat engine with 30% efficiency performs 2,500 J of work, how much heat would be discharged to the lower temperature reservoir?
    A. 8,333 J
    B. 5,833 J
    C. 2,500 J
    D. 3,333 J

Questions 22–24 refer to the following passage.

The particles that make up matter are always in motion and therefore have kinetic energy. Temperature is a measure of the average kinetic energy of the particles of a substance. The more the particles are in motion, the greater the temperature of the substance. **Heat** is a flow of energy from a higher temperature object to a lower temperature object. Heat is transferred by conduction, convection, or radiation.

Conduction is the transfer of heat or thermal energy resulting from differences in temperature between adjacent bodies or adjacent parts of a body. By this process, heat is transferred from one substance to another when the substances are in direct contact. Because the particles in solids are closer together than those in liquids or gases, solids generally conduct heat better than liquids or gases. An example of conduction is the transfer of heat to the handle of a pot sitting on a hot burner.

Convection is the transfer of heat from one place to another by the movement of fluids. As a fluid heats, the particles move farther apart and become less dense. The more dense fluid sinks and the warmer, less dense fluid rises.

Radiation is the transfer of energy by electromagnetic waves. It is a method of heat transfer that does not rely on any contact between the heat source and the heated object as is the case with conduction and convection. An example of radiation is the heat from the sun.

22. Kim put a cup of dry macaroni pasta into a pot of boiling water on the stove. After a minute, she noticed that the pieces of macaroni were rising and falling through the water. What type of heat transfer makes the macaroni do this?

A. conduction only
B. conduction and convection
C. radiation and conduction
D. convection only

23. How does heat get transferred from the hot end of a glass rod to the cold end?

A. The molecules from the hot end move to the cold end.
B. The molecules from the hot end send out radiation to the cold end.
C. The molecules from the hot end vibrate more and pass on the energy to the neighboring molecules.
D. The molecules from the hot end move from place to place, collide with the colder molecules, and pass on the energy to them.

24. Jim was conducting the following experiment: He placed a metal pot of water on a stove burner. He turned the stove burner on. After a few minutes, the air above the pot of water became hot. Describe how heat flows from the burner to the pot, through the water, and to the air above the pot.

Write your response on the lines.

Questions 25–27 refer to the following information.

An important effect of heat is that it can produce a change of state. Matter exists in any of the three states—solid, liquid, or gas. In a gas, the particles can move around freely and independently. In a liquid, particle movement is a bit constrained and limited to sliding and flow movement within its volume. In a solid, particle movement is restricted to only vibrational motion of particles in their fixed positions.

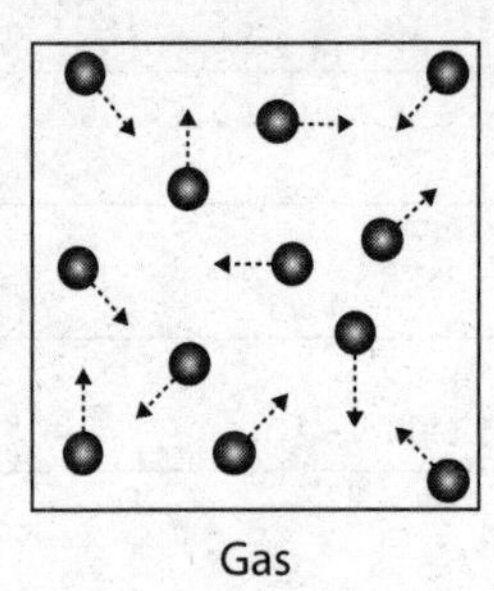
Gas

Liquid

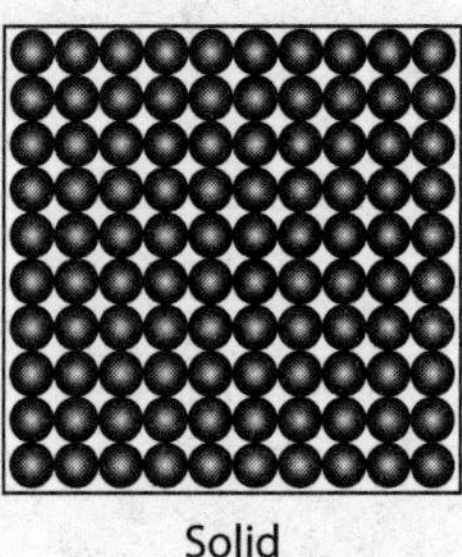
Solid

Thus a change of matter from one physical state to another is called change of state. The state of matter depends on its temperature and the pressure that is exerted on it.

There are six distinct changes of phase that happen to different substances at different temperatures:

- **Freezing**—A substance changes from a liquid to a solid.
- **Melting**—A substance changes from a solid to a liquid.
- **Condensation**—A substance changes from a gas to a liquid.
- **Evaporation or vaporization**—A substance changes from a liquid to a gas.
- **Sublimation**—A substance changes directly from a solid to a gas without going through the liquid state.
- **Deposition**—A substance changes directly from a gas to a solid without going through the liquid state.

25. In warm, humid weather, people who wear glasses experience a very common effect of phase change while getting off an air-conditioned bus: their glasses suddenly fog up. After a while, the glasses become clear again. Which of the following physical processes are involved in these phenomena?

    A. condensation followed by evaporation
    B. condensation followed by fusion
    C. solidification followed by evaporation
    D. solidification followed by fusion

Write your answers in the blanks.

26. During the change of state, when a solid changes into a liquid, the process is called ________________, and when a solid changes directly into a gas, the process is called ________________.

27. Which of the following is NOT a property of a gas?

    A. no definite shape
    B. can be compressed
    C. definite volume
    D. definite mass

Questions 28–31 refer to the following passage.

A chemical reaction is a process in which one or more substances, the reactants, are converted to form one or more chemically different substances, the products.

An **endothermic** reaction is a reaction that absorbs heat from its environment. Examples include the melting of ice under sunlight, or chemical reactions such as:

$$\mathbf{N_2(g) + O_2(g) + energy \rightarrow 2NO(g)}$$

An **exothermic** reaction is a reaction that releases heat. Examples include natural processes such as respiration and chemical reactions such as:

$$\mathbf{C(s) + O_2(g) \rightarrow CO_2(g) + energy}$$

Write your answer in the blank.

28. Ice cream melting on a hot sunny day is an example of an

 ________________ reaction.

*The following two questions contain two blanks, each marked "Select . . . ▼." Beneath each one is a set of choices. Indicate the choice that is correct and belongs in the blank. (**Note**: On the real GED test, the choices will appear as a "drop-down" menu. When you click on a choice, it will appear in the blank.)*

29. When quick lime (calcium oxide, or CaO) reacts with water, the reaction is considered to be exothermic. A student mixes calcium oxide and water in a test tube and touches its side surface. The tube becomes Select . . . ▼

Select . . . ▼
cold
hot
neutral

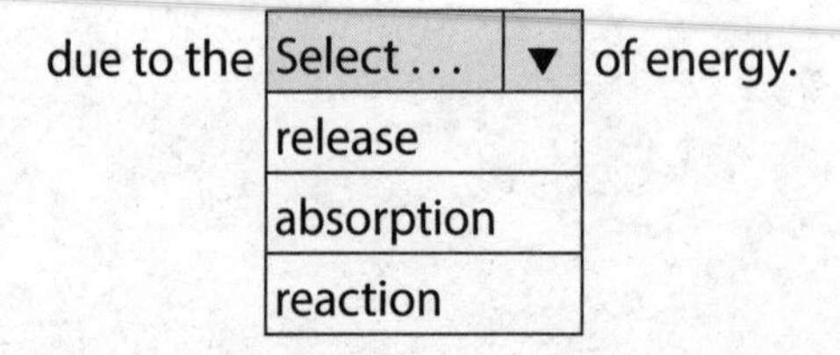

30. Consider the following reaction:

$H_3PO_4 + 3LiOH \rightarrow Li_3PO_4 + 3H_2O + energy$

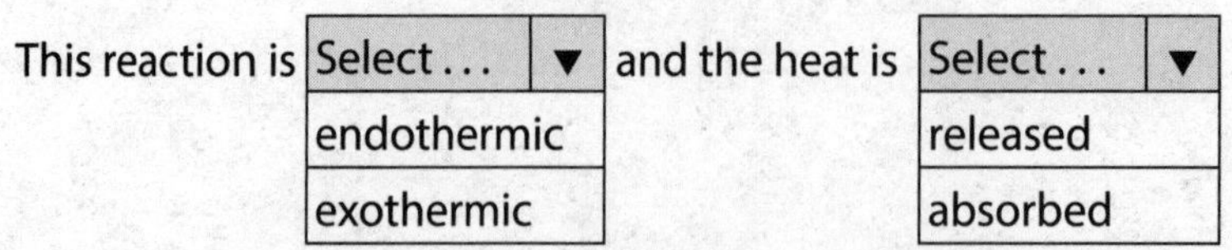

31. Explain why cellular respiration is considered an exothermic reaction.

Write your response on the lines.

Question 32 refers to the following passage.

Energy cannot be created or destroyed, but it can be transformed from one form to another. In the process of doing work, the object that is doing the work exchanges energy with the object on which the work is done. When work is done on an object, that object gains energy. The energy acquired by the objects on which work is done is known as mechanical energy. Mechanical energy is the energy that is possessed by an object due to its motion and/or due to its position. Thus the total amount of mechanical energy (T_{ME}) is the sum of the potential energy (P_E) and the kinetic energy (K_E) of an object, that is, $T_{ME} = P_E + K_E$, as shown in the graph.

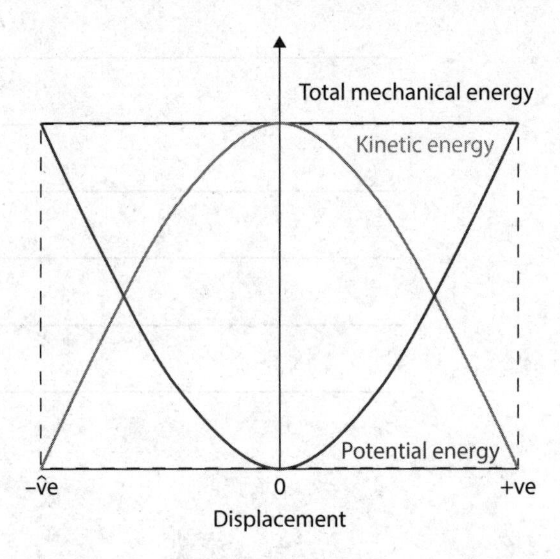

When only one type of force acts on an object when doing work, the total mechanical energy of that object remains constant. In such cases, the object's energy changes only its form. When an object itself is doing work, that object loses its potential energy, but that energy is regained in the form of kinetic energy and vice versa. Thus the total mechanical energy is always conserved.

Write your answer in the blank.

32. A projectile is launched from ground level. At the highest point in its trajectory, its total mechanical energy is ________________ its total mechanical energy at its launch position.

Questions 33–35 are based upon the following passage.

Fritz Haber was awarded the 1918 Nobel Prize in Chemistry for his work in creating the Haber process, a process by which ammonia is made. The reaction for the Haber process is as follows:

$$3H_2(g) + N_2(g) \leftrightarrow 2NH_3(l) + \text{heat}$$

33. Which of the following would NOT drive the reaction to make more ammonia?

    A. heating the reactants
    B. reacting the gases at a low pressure
    C. using a catalyst
    D. using higher concentrations of reactants

34. In order for the hydrogen and nitrogen to react, there must be

    A. effective collisions between the hydrogen and nitrogen gases so as to cause a rearrangement of atoms
    B. a sufficient amount of ammonia present to form the hydrogen and nitrogen
    C. contact between the hydrogen and nitrogen atoms
    D. smaller amounts of hydrogen and nitrogen present

35. If 12 grams of hydrogen gas are completely reacted with 56 grams of nitrogen gas to make ammonia, what is the total amount of ammonia that can be generated?

    A. 4.67 grams
    B. 0.214 grams
    C. 44 grams
    D. 68 grams

Questions 36–38 refer to the following passage.

Kinetic energy is the energy of motion. An object that has motion has kinetic energy. The amount of kinetic energy that an object in motion possesses depends on its mass (m) and speed (v). The following equation represents the kinetic energy (K_E) of an object

$$\textbf{Kinetic energy } (K_E) = \tfrac{1}{2} mv^2$$

where m = mass in kg of object; v = speed of object.

Kinetic energy is a scalar quantity; it does not have a direction. The kinetic energy of an object is completely described by its magnitude alone. The unit of measurement for kinetic energy is the joule.

36. What is the kinetic energy, in joules, of a 625-kg roller coaster car that is moving with a speed of 18.3 m/s?

    A. 1.05×10^4 J
    B. 1.05×10^5 J
    C. 2.05×10^5 J
    D. 1.06×10^4 J

37. Which form of energy does a flowing river possess?

    A. gravitational energy
    B. potential energy
    C. electrical energy
    D. kinetic energy

*The following question contains a blank marked "Select... ▼." Beneath it is a set of choices. Indicate the choice that is correct and belongs in the blank. (**Note:** On the real GED test, the choices will appear as a "drop-down" menu. When you click on a choice, it will appear in the blank.)*

38. If the speed of an object is doubled, then its kinetic energy is

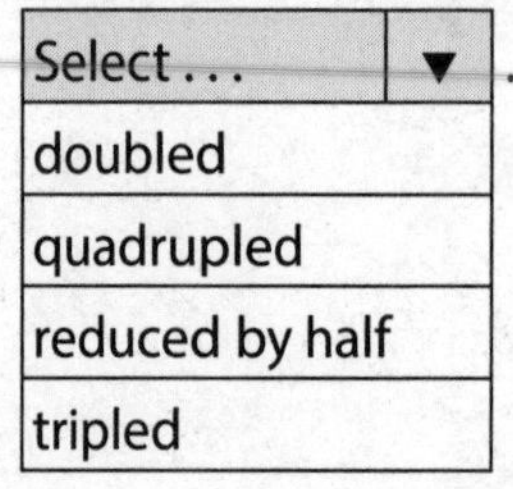

39. Explain how a hammer utilizes mechanical energy to do work when it hits a nail.

Write your response on the lines.

____

____

____

____

____

____

____

____

____

____

40. The photoelectric effect and the double slit experiment are two methods for demonstrating the dual nature of light. They prove that light can act like both

A. a wave and a particle
B. a fusion reaction and a fission reaction
C. matter and anti-matter
D. protons and neutrons

Questions 41 and 42 refer to the following passage.

Energy exists in many forms around us. The development of modern civilization has been accomplished because scientists and engineers had learned the mechanism of converting one form of energy into another to do useful work. The transformation of energy from coal into steam and then into mechanical energy in engines that do heavy work enabled the Industrial Revolution.

There are different forms of energy around us, such as mechanical energy (potential energy and kinetic energy), chemical energy, electrical energy, radiant energy, nuclear energy, and thermal energy. All these forms of energy are related and can be transformed or changed into the other forms to do useful work.

41. Hydroelectric dams make use of the differences in water levels to generate electricity. Which of the following shows the energy conversion in a dam?

    A. Gravitational potential energy → Kinetic energy → Electrical energy
    B. Electrical energy → Heat energy → Electrical energy
    C. Kinetic energy → Heat energy → Electrical energy
    D. Kinetic energy → Heat energy → Light energy

42. Which of the following shows the energy conversions that are required to illuminate a battery-operated table lamp?

    A. Electrical energy → Chemical energy → Light energy
    B. Electrical energy → Chemical energy → Kinetic energy
    C. Chemical energy → Electrical energy → Light energy
    D. Chemical energy → Electrical energy → Kinetic energy

Questions 43–48 refer to the following chart.

Energy Sources

Nonrenewable Energy Sources: Energy resources that cannot be replenished.

Uranium - Nuclear Energy:

- This is obtained by nuclear fission, which is the process of splitting apart uranium atoms in a controlled manner that creates energy.
- This is quite dangerous because if the chain reaction of splitting the atoms is not controlled very carefully, an atomic explosion could occur.
- The fission process gives off heat energy, which is used to boil water in a power plant's reactor core. The steam created with this water is used to turn a turbine, generating electricity.

Fossil fuel: Coal, oil, and gas are called "fossil fuels" because they have been formed from the organic remains of prehistoric plants and animals.

Petroleum/Oil:

- Petroleum is made when organic matter settles in water that has lost its dissolved oxygen and is then compressed under immense heat and pressure for millions of years.
- It is extracted and turned into a variety of fuel sources including petrol or gasoline, diesel, propane, jet fuel, heating oil, and paraffin wax.
- It is also called crude oil.

Natural Gas:

- This is a by-product of decomposition.
- The most common form of natural gas is methane, created from decays of organic matter.
- Natural gas burns cleaner than oil or coal, releases less carbon dioxide, causes less pollution, and is environment friendly, which has encouraged its use.

Coal:

- It is composed of organic matter that decomposed and then formed into carbon rock under immense pressure.
- Coal is generally highly combustible and the world's most-used resource for electrical generation.
- Burning coal releases massive amounts of carbon dioxide into the atmosphere, which is the primary factor in the greenhouse effect.

Renewable Energy Sources: Energy that comes from a source that's constantly renewed or can be replenished naturally in a short period of time.

Solar: This form of energy relies on the nuclear fusion power from the core of the sun. This energy can be collected and converted in a few different ways. For example, solar water heating with solar panels.

Wind Power: The movement of the atmosphere is driven by differences of temperature at the Earth's surface due to varying temperatures of the Earth's surface when heated by sunlight. For example, wind energy can be used to pump water or generate electricity.

Hydroelectric: This form uses the gravitational potential of elevated water, for example, hydroelectric dams.

Biomass:

- It is the term for energy from plants.
- Energy in this form is very commonly used throughout the world.
- Unfortunately, the most popular is the burning of trees for cooking and warmth. This process releases copious amounts of carbon dioxide gases into the atmosphere and is a major contributor to air pollution.
- Some of the more modern forms of biomass energy are methane generation and production of alcohol for automobile fuel and fueling electric power plants.

Hydrogen Fuel Cells:

- These are not strictly renewable energy resources but are very abundant in availability and are very low in pollution when utilized.
- Hydrogen can be burned as a fuel, typically in a vehicle, with only water as the combustion product. This clean burning fuel can mean a significant reduction of pollution in cities.

Geothermal Power:

- This is obtained from the energy of naturally occurring hot water and steam that is heated underground.
- This energy can be used to generate electricity. This possibility is limited by geography.

Write your answer in the blank.

43. The nonrenewable energy source that has the lowest emissions is ________________.

44. What are fossil fuels?

Write your response on the lines.

____

____

____

____

____

____

____

____

____

____

45. Natural gas
    - A. is a primary component in octane
    - B. produces less greenhouse gases per energy unit than coal
    - C. will be depleted by 2030 at current usage rates
    - D. is transported primarily by truck and rail

*The following question contains a blank marked "Select . . . ▼." Beneath it is a set of choices. Indicate the choice that is correct and belongs in the blank. (**Note:** On the real GED test, the choices will appear as a "drop-down" menu. When you click on a choice, it will appear in the blank.)*

46. The energy that is NOT derived from the sun is 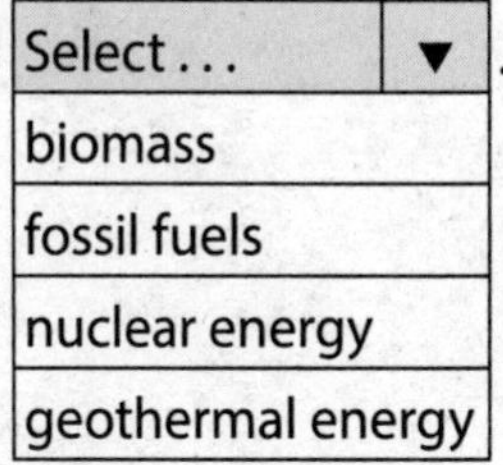

.

47. Indicate the box in which each of the following terms belongs.

- Volcanos
- Trees
- Crops
- Hot springs
- Plants
- Natural geysers

A: Geothermal sources	B: Biomass sources

*The following question contains a blank marked "Select . . . ▼." Beneath it is a set of choices. Indicate the choice that is correct and belongs in the blank. (**Note:** On the real GED test, the choices will appear as a "drop-down" menu. When you click on a choice, it will appear in the blank.)*

48. Nuclear Select . . . ▼ is the process that is responsible for the sun's

fission
fusion

radiant energy, commonly known as solar energy.

Questions 49 and 50 refer to the following chart.

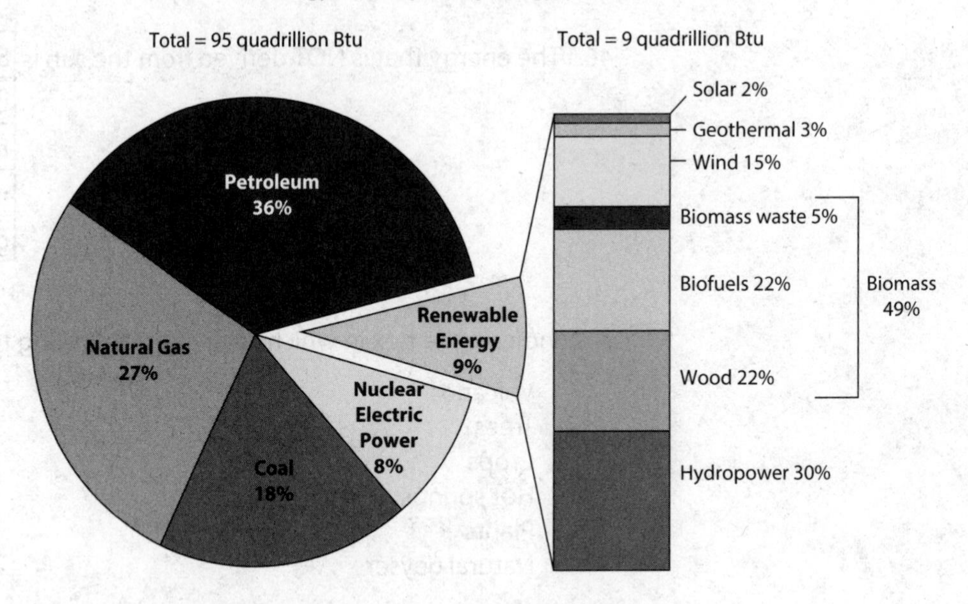

Note: Sum of components may not qual 100% due to independent rounding.
Fossil fuel comprises coal, petroleum, and natural gas.

*The following question contains a blank marked "Select . . . ▼." Beneath it is a set of choices. Indicate the choice that is correct and belongs in the blank. (**Note:** On the real GED test, the choices will appear as a "drop-down" menu. When you click on a choice, it will appear in the blank.)*

49. According to the energy consumption report of 2012 by the US Energy Information Administration, the most utilized renewable energy resource in the United States is Select . . . ▼.

biofuel
petroleum
hydropower
biomass

50. According to the energy consumption report of 2012 by the US Energy Information Administration, what percentage of energy used in the United States is from fossil fuels?

A. 36%
B. 27%
C. 60%
D. 81%

51. Read the following facts:

    Fact 1: When sunlight is absorbed by an object, the object becomes warmer.

    Fact 2: Dark-colored objects absorb more sunlight than light-colored objects.

    On a hot, sunny day, how can you keep coolest?

    A. by wearing dark-colored clothes and staying out of direct sunlight
    B. by wearing light-colored clothes and staying in direct sunlight
    C. by wearing dark-colored clothes and staying in direct sunlight
    D. by wearing light-colored clothes and staying out of direct sunlight

Questions 52 and 53 refer to the following information.

The greenhouse effect is a natural process that warms Earth's surface. When the sun's energy reaches Earth's atmosphere, some of it is reflected back to space and the rest is absorbed and re-radiated by greenhouse gases.

52. What is the relationship between fossil fuels and the greenhouse effect?

    A. Burning fossil fuels decreases incoming solar radiation.
    B. Burning fossil fuels decreases the absorption capacity of greenhouse gases.
    C. Burning fossil fuels lowers the greenhouse effect.
    D. Burning fossil fuels releases carbon dioxide, which is a greenhouse gas.

53. How can aerosol sprays and refrigerator coolants containing CFC contribute to global warming?

    A. The liquids that are released cause more specific heat to be trapped in the atmosphere.
    B. The gases that are released directly heat up Earth's surface.
    C. The liquids that are released cause Earth's atmosphere and surface to heat up.
    D. The gases that are released cause more specific heat to be trapped in the atmosphere.

Questions 54–59 refer to the following passage.

Waves are disturbances that propagate energy through a medium. Propagation of the energy depends on interactions between the particles that make up the medium. Particles move as the waves travel, but there is no net motion of particles. Thus, mechanical waves require matter through which to propagate. These waves are of the following types:

- **Longitudinal waves**—Movement of the particles is parallel to the direction the wave travels; that is, the direction of displacement is the same as the direction of propagation. Sound waves are longitudinal waves.
- **Transverse waves**—Movement of the particles is perpendicular to the direction the wave travels; that is, the direction of displacement is perpendicular to the direction of propagation. Movement of a wave through a stretched rope or a trampoline is an example of this type of wave.

Wavelength (λ) is the distance between identical points in the adjacent waves such as from crest to crest or trough to trough. It is measured in meters.

Frequency (f) describes the number of waves that passes a fixed place in a given amount of time. It is measured in hertz or cycles/s.

A **period** (T) is the time needed for one complete cycle of vibration to pass through a given point.

Frequency (f) and period (T) have a reciprocal relationship that can be expressed mathematically as

$$T = \text{total time / number of cycles}$$

$$\Rightarrow T = 1 / \text{frequency} = 1/f$$

The wave equation states the mathematical relationship between the speed (v) of a wave and its wavelength (λ) and frequency (f).

$$\text{speed} = \text{wavelength} \times \text{frequency}$$

$$\Rightarrow v = \lambda \times f$$

or

$$v = f/T \text{ since, } \lambda = 1/f$$

The amplitude (A) of a wave refers to the maximum amount of displacement of a particle of the medium from its rest position.

54. A swing completes a back-and-forth cycle every 2 seconds. What is the frequency of the swing?

    A. 0.5 Hz
    B. 1 Hz
    C. 0.2 Hz
    D. 2 Hz

55. Two anchored boats bob up and down, returning to the same up position every 3 seconds. When one is up, the other is down.

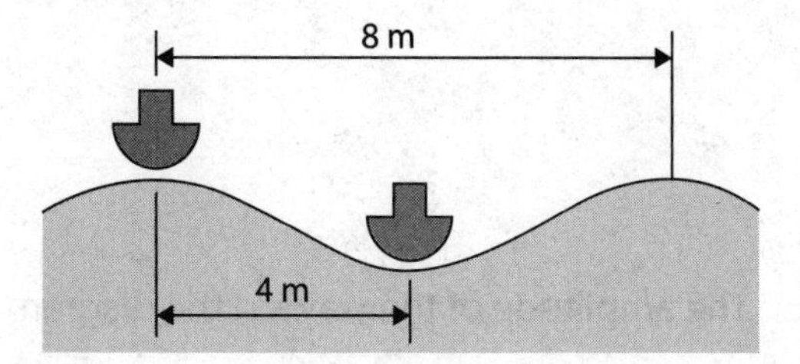

    There are never any wave crests between the boats. Calculate the speed of the waves.

    A. 4.67 m/s
    B. 2.67 m/s
    C. 2.47 m/s
    D. 4.47 m/s

56. A transverse wave is transporting energy from east to west. The particles of the medium will move in which direction?

    A. east to west only
    B. east to west and west to east
    C. north to south only
    D. north to south and south to north

*The following question contains two blanks, each marked "Select . . . ▼." Beneath each one is a set of choices. Indicate the choice that is correct and belongs in the blank. (**Note:** On the real GED test, the choices will appear as a "drop-down" menu. When you click on a choice, it will appear in the blank.)*

57. Consider the following diagram:

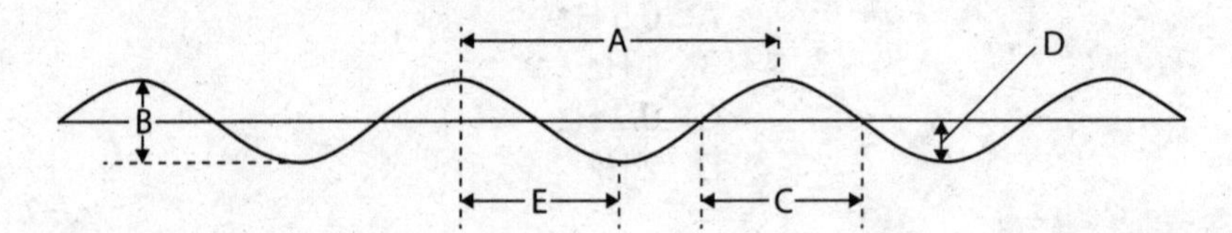

The wavelength of the wave in the diagram is given by the letter [Select . . . ▼].

Select . . . ▼
A
E
C
D

The amplitude of the wave in the diagram is given by the letter [Select . . . ▼].

Select . . . ▼
A
C
E
D

58. A supermarket shopper approaches the exit door with a cart full of groceries. When the front of the cart approaches the door, a small red light on a sensor above the door turns on. The door then opens. This demonstrates

A. diffraction of light
B. refraction of light
C. the photoelectric effect
D. the destructive interference of light waves

59. Indicate the box in which each of the following terms belongs.

- Sound waves
- Light waves
- Crest
- Compression
- Trough
- Rarefaction

A: Longitudinal waves	B: Transverse waves

60. Which particles define what element a particular atom is?

A. nucleons
B. protons
C. neutrons
D. electrons

Questions 61 and 62 are based on the following information.

Energy is needed to break the bonds between the atoms within molecules and the bonds that exist between molecules. The energy required to break these bonds can vary depending on the types of bonds involved and the types of elements involved. The chart below shows the melting points for substances A, B, C, and D:

Substance	**A**	**B**	**C**	**D**
Melting Point (°C)	50	950	1350	1860

61. Which substance contains the strongest bonds?

A. A
B. B
C. C
D. D

62. Substance A is most likely made of which type(s) of elements?

A. metals
B. a metal and a nonmetal
C. nonmetals
D. a semimetal and a nonmetal

Questions 63–65 refer to the following passage.

Speed (v) is defined as the rate at which an object covers a distance (d).

speed (v) = distance (d) / time (t), that is,

$$v = d/t$$

Displacement is the difference in the distance traveled by an object from the initial to the final position.

Velocity ($\mathbf{v}$) is a vector quantity that refers to "the rate at which an object changes its position."

$$\mathbf{v} = \textbf{displacement / time}$$

Velocity differs from speed in that it is a vector quantity, while speed is a scalar quantity. Velocity is speed with a direction. Therefore, positive and negative velocities are simply velocities in opposite directions.

Question 63 is based on the following diagram. Write your answers in the blanks.

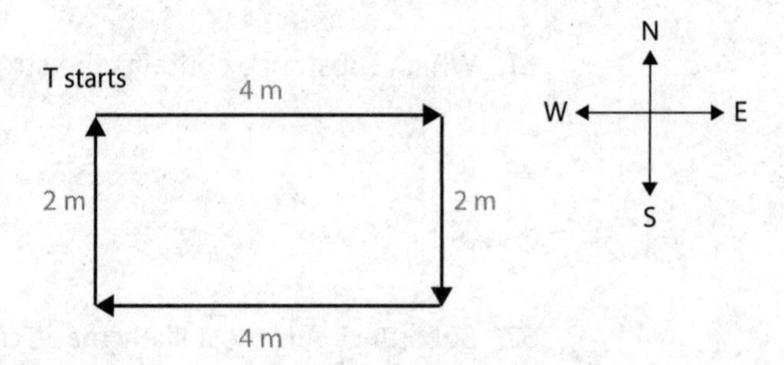

63. According to the diagram, a physics teacher (T) starts from a point and walks following the arrowed path and the distance and direction as shown. The entire motion lasts for 24 seconds. The speed of the teacher is _______________, and the velocity of the teacher is _______________.

64. If a cyclist in the Tour de France travels southwest a total distance of 12,250 m in 1 hour, what would the velocity of the cyclist be?

   A. 3.80 m/s southwest
   B. 3.40 m/s southwest
   C. 3.40 m/s northwest
   D. 3.80 m/s southeast

*The following question contains a blank marked "Select . . . ▼." Beneath it is a set of choices. Indicate the choice that is correct and belongs in the blank. (**Note:** On the real GED test, the choices will appear as a "drop-down" menu. When you click on a choice, it will appear in the blank.)*

65. The initial velocity of a body starting from rest is Select . . . ▼ .

Select . . . ▼
1 m/s
2 m/s
zero

Questions 66–69 refer to the following passage.

Momentum is the quantity of motion of a moving body, measured as a product of its mass and velocity:

$$\textbf{momentum } (p) = \textbf{mass } (m) \times \textbf{velocity } (v)$$

The unit of momentum is kg·(m/s). Momentum is a vector quantity as it has both magnitude and direction. Momentum of an object is always in the same direction as its velocity.

66. A 1-kg ball traveling forward at a velocity of 4.0 m/s strikes a wall and bounces straight backward at a velocity of −2.0 m/s. What is the momentum of the ball?

    A. +8 kg·(m/s)
    B. −8 kg·(m/s)
    C. −6 kg·(m/s)
    D. +6 kg·(m/s)

*The following question contains a blank marked "Select . . . ▼." Beneath it is a set of choices. Indicate the choice that is correct and belongs in the blank. (**Note:** On the real GED test, the choices will appear as a "drop-down" menu. When you click on a choice, it will appear in the blank.)*

67. A bullet of mass 4 g moving with a velocity of 400 m/s will have

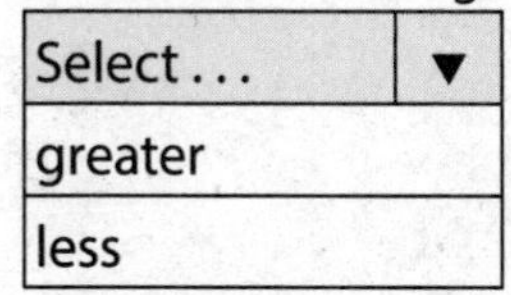

momentum than a baseball of mass 150 g thrown with a speed of 90 km/h.

68. Which of the following statements is true based on the equation for momentum?

    A. If mass gets doubled, momentum will get halved.
    B. If velocity gets doubled, momentum will get halved.
    C. If both mass and velocity get doubled, momentum will increase by four times.
    D. If both mass and velocity get halved, momentum will increase by four times.

69. When fighting fires, a firefighter must use great caution to hold a hose that emits large amounts of water at high speeds. Why would such a task be difficult?

Write your response on the lines.

____

____

____

____

____

____

____

____

____

____

Question 70 refers to the following information.

The principle of conservation of momentum states that for a collision occurring between two objects, provided that no external forces act on the system, the total momentum before the collision is equal to the total momentum after the collision:

$$[(m_1 \times v_{1a}) + (m_2 \times v_{2a})]_{before} =$$
$$(m_1 \times v_{1b}) + (m_2 \times v_{2b})]_{after}$$

where m_1 = mass of object 1; m_2 = mass of object 2; v_{1a} = velocity of object 1 before collision; v_{2a} = velocity of object 2 before collision; v_{1b} = velocity of object 1 after collision; and v_{2b} = velocity of object 2 after collision.

70. Consider the following case:

Before collision:

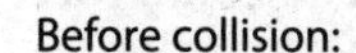

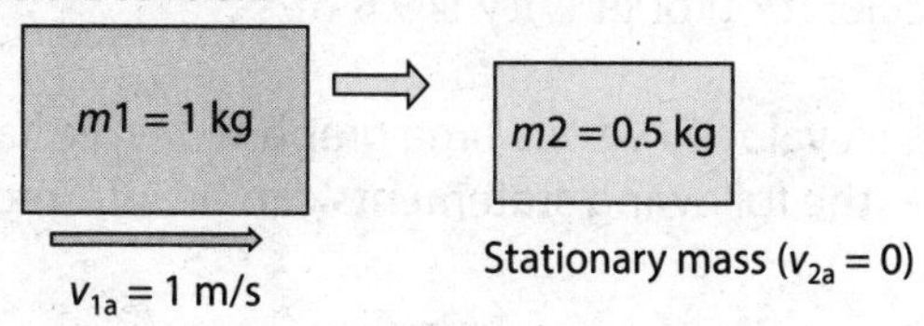

v_{1a} = 1 m/s

After collision:

m1 = 1 kg

m2 = 0.5 kg

v_{1b} = ?

v_{2b} = 1.33 m/s

After the collision, how fast is the first mass moving?

A. 0.335 m/s
B. 0.035 m/s
C. 1.335 m/s
D. 335 m/s

Questions 71–73 refer to the following information.

Acceleration (a) is the rate of change of velocity (v) of an object; that is,

$$\textbf{acceleration} = \textbf{velocity/time}$$

or

$$a = v/t$$

The unit of acceleration is m/s2. Acceleration is a vector quantity because it has both magnitude and direction. The direction of the acceleration vector depends on two things: whether the object is speeding up or slowing down, and whether the object is moving in the positive or negative direction. The general principle for determining acceleration is that if an object is slowing down, then its acceleration is in the opposite direction of its motion.

The acceleration of gravity (g) is the acceleration for any object moving under the sole influence of gravitational force. The numerical value for the acceleration of gravity is 9.8 m/s^2.

71. A velocity versus time graph for a free-falling object is shown. Which of the following statements can be inferred from this graph?

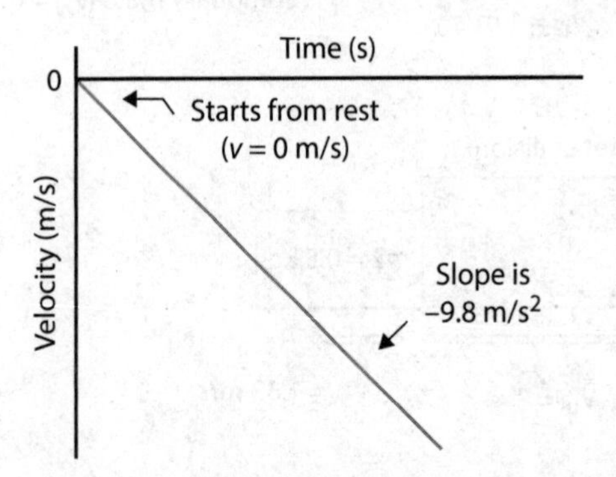

A. It shows negative velocity and positive acceleration.
B. It shows negative velocity and negative acceleration.
C. It shows negative velocity and positive momentum.
D. It shows a body at rest.

72. Under which condition is an object accelerating?

A. only when its speed changes
B. only when its direction changes
C. when its speed or direction changes
D. only when its velocity increases

73. A pitcher requires about 0.2 second to throw a baseball. If the ball leaves his hand with a velocity of 80 m/s, what is its acceleration?

A. 40 m/s^2
B. 400 m/s^2
C. 40 m/s
D. 4 m/s

Questions 74–78 refer to the following passage.

Newton's first law of motion states that an object continues to be in a state of rest or motion unless it is acted on by an external force. The net force acting on the body is the total of all the forces acting on the body. Newton's first law is sometimes called the law of inertia, which is the tendency of an object to resist changes in velocity. Inertia of an object is directly proportional to its mass.

74. A book is placed on a dashboard of a car that is stopped at a traffic light. As the car starts to move forward, the book slides off the dashboard. Pick the most correct explanation for the book sliding off the dashboard.

    A. There is grease on the dashboard, so the book slides off the dashboard.
    B. Due to inertia, the book slides off the dashboard.
    C. The book is heavy, so it slides off the dashboard.
    D. Air resistance made the book move backward.

75. Matthew is playing putt-putt golf. The fifteenth hole at the golf course has a large metal rim that putters must use to guide their ball toward the hole. Matthew guides a golf ball around the metal rim as shown in the diagram:

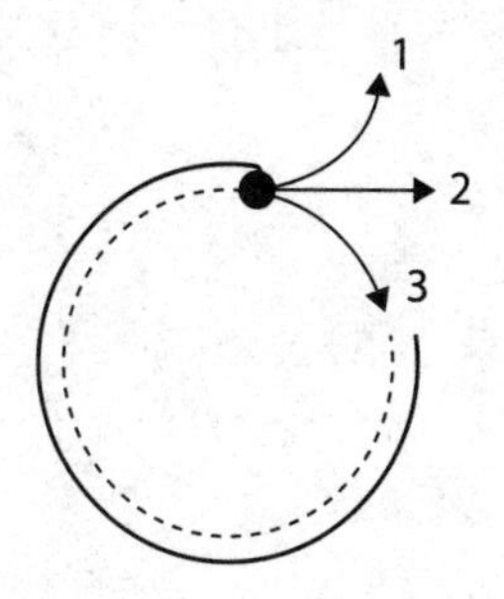

    When the ball leaves the rim, which path, 1, 2, or 3, will the golf ball follow? Explain your answer.

Write your response on the lines.

____

____

____

____

____

____

76. What is the net force acting on the object shown in the diagram?

A. 25 N to the right
B. 0 N
C. 20 N to the right
D. 20 N to the left

77. A distance–time graph plots distance marked against regular intervals of time.

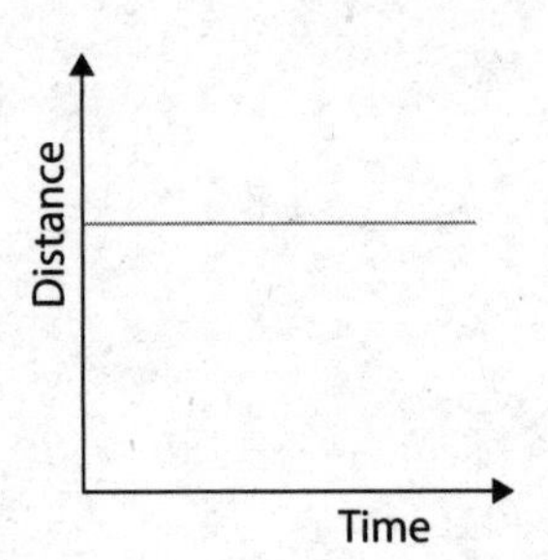

Which of the following can be inferred from the graph?

A. The object is at rest.
B. The object has a net force to the right acting on it.
C. The object has positive velocity.
D. The object has positive momentum.

78. Which of the following is the best example of Newton's first law of motion?

A. Whenever you push against a wall, the wall pushes back.
B. When a bat hits a ball, the ball hits the bat.
C. A rider on a running horse is thrown forward when the horse stops suddenly.
D. A 100-g baseball will accelerate more slowly than a 50-g baseball because of its greater mass.

Questions 79–82 refer to the following passage.

Newton's second law is given by the equation

$$a = F/m$$

where a is the acceleration of an object in m/s^2, F is the force applied to the object, and m is the mass of the object being accelerated in kilograms. Recall that acceleration is the rate of change of velocity of an object; that is,

$$a = v/t$$

79. A soccer ball weighing 500 g is lying near the center line of the field. A player kicks it with a force of 400 N. What is the velocity of the ball after 5 seconds?

    A. 4,000 m/s
    B. 3,200 m/s
    C. 1,600 m/s
    D. 800 m/s

80. A driver accelerates his car first at the rate of 1.8 m/s^2 and then at the rate of 1.2 m/s^2. What will be the ratio of the two forces exerted by the engine at the two rates?

    A. 1:2
    B. 2:1
    C. 2:3
    D. 3:2

Write your answer in the blank.

81. If the net force acting on a sliding block is tripled, the acceleration will increase by _______________ times.

*The following question contains a blank marked "Select . . . ▼." Beneath it is a set of choices. Indicate the choice that is correct and belongs in the blank. (**Note**: On the real GED test, the choices will appear as a "drop-down" menu. When you click on a choice, it will appear in the blank.)*

82. A train with cars of equal masses is moving with constant velocity along a level section of track. The net force on the first car is Select . . . ▼ the net force on the last car.

Select . . . ▼
equal to
greater than
less than
opposite to

Questions 83 and 84 refer to the following passage.

Newton's third law states that all forces come in pairs. The two forces in the pair are equal in strength and opposite in direction.

83. Based on Newton's third law of motion, identify the action–reaction force pair involved when a diver jumps off a diving board.

Write your response on the lines.

____

____

____

____

____

____

____

____

*The following question contains a blank marked "Select . . . ▼." Beneath it is a set of choices. Indicate the choice that is correct and belongs in the blank. (**Note**: On the real GED test, the choices will appear as a "drop-down" menu. When you click on a choice, it will appear in the blank.)*

84. The action and reaction forces referred to in Newton's third law must act on Select . . . ▼ objects.

Select . . . ▼
identical
different
equivalent
similar

85. Ernest Rutherford's gold foil experiment proved that atoms are mostly empty space. His experiment also helped to determine that the mass of an atom is highly concentrated in the nucleus of the atom. Which particle accounts for the least part of the mass of an atom?

A. nucleon
B. proton
C. neutron
D. electron

Questions 86–88 refer to the following passage.

When a force acts on an object to cause a displacement (d) of the object, it is said that work (W) was done on the object.

Power is the rate at which work is done. It is the work/time ratio. Mathematically, it is computed using the following equation:

$$\textbf{power } (P) = \textbf{work } (W) \,/\, \textbf{time } (t)$$
$$\Rightarrow \quad P = [\textbf{force } (F) \times \textbf{displacement } (dx)] \,/\, t$$
$$\Rightarrow \quad P = F \times \textbf{velocity}$$

86. A man does a push-up by applying a force, which results in 0.50 J of work. If it takes 2 seconds to do the push-up, what is his power?

A. 0.50 W
B. 0.25 W
C. 1.25 W
D. 0.025 W

87. Use your understanding of work and power to answer the following question. Two teammates, Randy and Aaron, are in the weight-lifting room. Randy lifts the 100-pound barbell over his head 10 times in 5 minutes; Aaron lifts the same 100-pound barbell over his head 10 times in 3 minutes. Which teammate does the most work, and who delivers the most power and why?

Write your response on the lines.

88. The power of a motor pump is 2 kW. How much water per minute can the pump raise to a height of 10 m? (Under the influence of gravitational force, work = mass (*m*) × gravitational field strength (*g*) × height (*h*), where $g = 10\ \text{ms}^{-2}$.)

    A. 1,200 kg
    B. 1,200 g
    C. 120 kg
    D. 120 kg

Questions 89 and 90 refer to the following passage, illustration, and graph.

A block slides without friction down a slick ramp. The graph shows the values of both the potential energy and the kinetic energy of the block as it slides. *d* is the point along the ground that is directly below the sliding block. The value of *d* changes from 0 to 10.

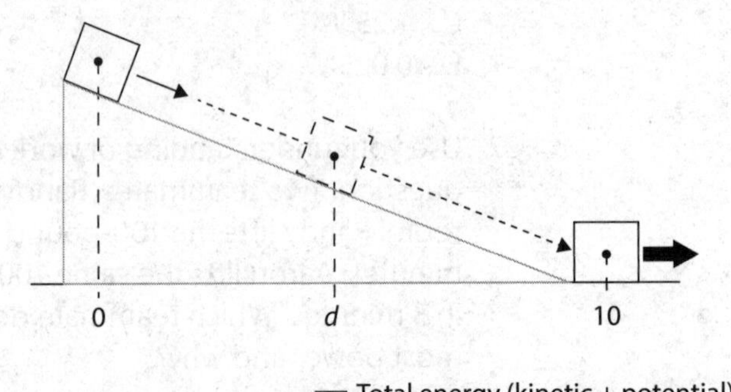

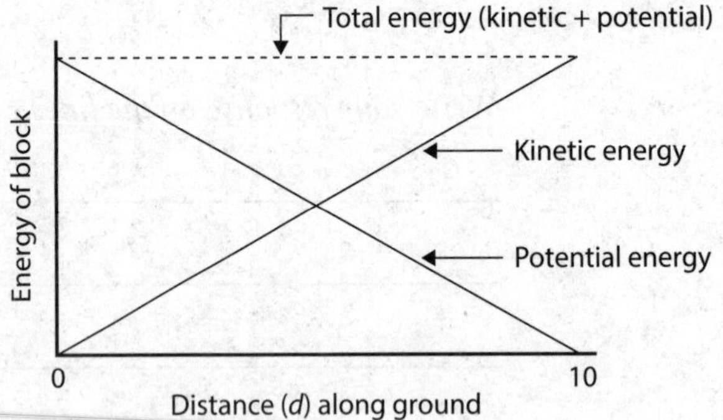

89. At what value of *d* is the kinetic energy of the block about equal to its potential energy?

    A. $d = 0$
    B. $d = 2$
    C. $d = 5$
    D. $d = 7$

90. Which statement about the block is NOT true?

    A. The total energy does not change.
    B. All of the block's potential energy is changed to kinetic energy.
    C. When $d = 0$, the kinetic energy is 0.
    D. All of the block's kinetic energy is changed to potential energy.

Questions 91–94 refer to the following information.

Simple machines are machines that do work with only one movement. They affect the way work is done. They can increase speed, change the direction of a force, or increase force. The following chart lists and defines some simple machines.

Simple Machine	Brief Definition
Lever	A bar that pivots about a fixed point
Wheel and axle	An arrangement of a wheel (or crank) and an axle that turn on the same axis, used to amplify force
Pulley	An arrangement of rope and pulleys that allows the operator to trade force for distance; a block and tackle
Inclined plane	A flat surface tilted at an angle, like a ramp, used for lifting a load that would be too heavy to lift straight up; the load must travel a greater distance, but pushing it requires less force
Wedge	A triangular-shaped tool used to separate two objects easily
Screw	A bolt with a helical ridge that amplifies force by a small rotational force

*The following question contains a blank marked "Select... ▼." Beneath it is a set of choices. Indicate the choice that is correct and belongs in the blank. (**Note:** On the real GED test, the choices will appear as a "drop-down" menu. When you click on a choice, it will appear in the blank.)*

91. A Select... ▼ is an example of a wedge.

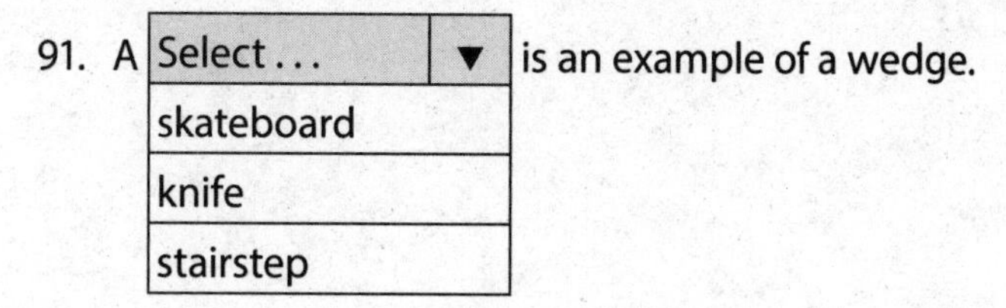

92. Which is an example of someone using a simple machine to do work?

    A. A boy runs across a football field.
    B. A banker counts money.
    C. A mother pushes a stroller up a ramp into a building.
    D. A girl eats a sandwich.

*The following question contains a blank marked "Select . . . ▼." Beneath it is a set of choices. Indicate the choice that is correct and belongs in the blank. (**Note:** On the real GED test, the choices will appear as a "drop-down" menu. When you click on a choice, it will appear in the blank.)*

93. Vivian is using a screwdriver to insert a screw. The screwdriver functions as a Select . . . ▼ .

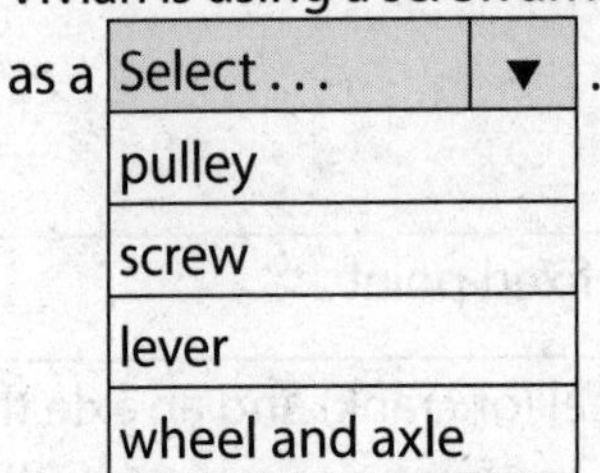

94. Which phrase best describes the principle involved in using an inclined plane?

A. greater force, shorter distance
B. greater force, longer distance
C. less force, shorter distance
D. less force, longer distance

Questions 95 and 96 refer to the following information.

A simple machine uses a single applied force to do work on a load. Ignoring friction losses, the work done by the applied force is equal to the work done on the load. However, the machine can amplify the load force (or output force) by trading off force against the distance traveled by the load. The ratio of the output force to the applied force is called the mechanical advantage. The mechanical advantage calculation for different simple machines is shown in the following table.

Simple Machine	Mechanical Advantage Calculation
Lever	The ratio of the length of the effort arm of the lever to the length of the load arm of the lever
Wheel and axle	The ratio of the radius of the wheel to the radius of the axle
Pulley (pulley system)	The total number of sections of rope pulling up on the object
Inclined plane	The length of the inclined plane divided by the height of the inclined plane (mechanical advantage = slope length/height)
Wedge	The length of either slope divided by the thickness of the big end
Screw	The circumference of the screw divided by the pitch of the screw

95. What is the mechanical advantage of a lever with an effort arm of 1 m and a load arm of 20 cm?

A. 5
B. 10
C. 20
D. 100

Write your answer in the blank.

96. Cassie uses a ramp to load a heavy box from the street into a truck. The bed of the truck is 1.26 m above the street. If the ramp is 5.3 m long, what is the mechanical advantage of the ramp (ignoring friction)? Round your answer to the nearest tenth: ________________

Questions 97–100 refer to the following information.

Matter is anything that occupies space and has mass. Matter can be a solid such as steel, a liquid such as water, or a gas such as air. Elements are pure substances composed of only one type of atom. An atom is the smallest particle of an element that has the properties of the element. An atom can be modeled by the diagram below.

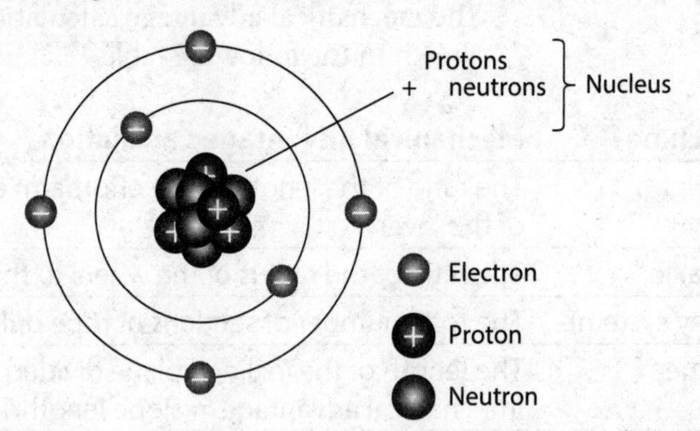

Neutral atoms have a charge of zero. If an atom gains electrons, it becomes negatively charged, and if an atom loses electrons, it becomes positively charged. Once an atom has an electrical charge, it is called an ion. The number of protons never changes in an atom and is unique for every element.

97. Which of the following has a positive charge?

    A. proton
    B. neutron
    C. atom
    D. electron

Write your answers in the blanks.

98. In a neutral atom, the number of ________________ equals the number of ________________.

*The following question contains a blank marked "Select . . . ▼." Beneath it is a set of choices. Indicate the choice that is correct and belongs in the blank. (**Note:** On the real GED test, the choices will appear as a "drop-down" menu. When you click on a choice, it will appear in the blank.)*

99. The subatomic particle that distinguishes one element from another is a(n) Select . . . ▼ .

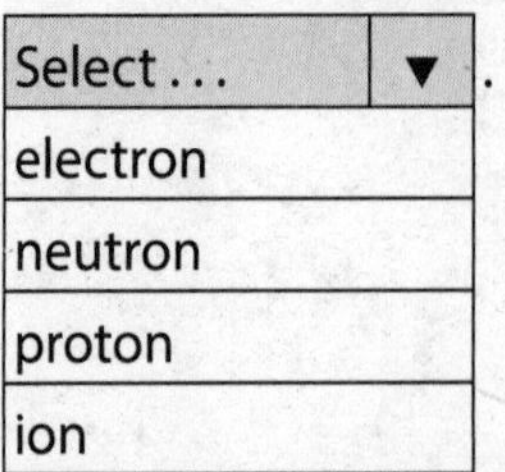

Select . . . ▼
electron
neutron
proton
ion

100. Magnesium has 12 protons and electrons. When charged, it loses 2 electrons. Which symbol represents the charged magnesium (Mg) ion?

A. Mg^{2-}
B. Mg
C. Mg^{2+}
D. Mg^{-}

Questions 101 and 102 are based on the following information.

Elements are pure substances composed of only one type of atom. For example, hydrogen, carbon, and oxygen are elements. Each element has its own atomic number. The **atomic number** is equal to the number of protons in the nucleus of a single atom and determines the identity of the element. For example, every hydrogen atom has one proton in the nucleus, and thus the atomic number of hydrogen is 1. A neutral atom contains an equal number of electrons and protons.

Every atom of a given element contains the same number of protons. However, different atoms of the same element can contain different numbers of neutrons. An atom's mass number is the total number of protons and neutrons in the nucleus of that atom.

*The following question contains a blank marked "Select... ▼." Beneath it is a set of choices. Indicate the choice that is correct and belongs in the blank. (**Note:** On the real GED test, the choices will appear as a "drop-down" menu. When you click on a choice, it will appear in the blank.)*

101. The atomic number for an atom containing 20 protons, 21 neutrons, and 20 electrons is.

Select... ▼
20
21
40

102. Carbon has an atomic number of 6. Which of the following arrangements describes a neutral carbon atom?

A. The total number of protons is 6 and the total number of electrons is 8.
B. The total number of protons is 6 and the total number of electrons is 7.
C. The total number of protons is 6 and the total number of electrons is 6.
D. The total number of protons is 6 and the total number of electrons is 10.

Questions 103 and 104 are based on the following information.

Atoms of the same element have the same number of protons, but they can have different numbers of neutrons and therefore different masses. Atoms with the same atomic number (the number of protons) but with different masses are called **isotopes**. Isotopes are identified by the total number of protons and neutrons in the nucleus. For example, the table below lists the isotopes of the element oxygen.

Isotopes of Oxygen			
	^{16}O	^{17}O	^{18}O
Number of protons	8	8	8
Number of neutrons	8	9	10

103. An isotope of argon (atomic number = 18) has a mass number of 40. How many neutrons are in this isotope?

A. 18
B. 40
C. 22
D. 58

104. Which one of the following pairs are isotopes?

A. O_2 and O_3
B. ^{35}Cl and ^{37}Cl
C. $I_2(g)$ and $I_2(s)$
D. F^+ and F^-

Questions 105 and 106 are based on the following information.

Helium is a stable element that will not react or burn with oxygen in air, and the low density of helium gas allows it to float easily. That is why helium gas is used to safely fill balloons. A balloon filled with hydrogen will float in air just like a balloon filled with helium. However, a balloon filled with hydrogen is dangerous because hydrogen can burn in oxygen. The reaction of the burning of hydrogen is as follows:

$$2H_2(g) + O_2(g) \rightarrow 2H_2O(l) + \text{heat}$$

105. This reaction can be classified as a(n)

A. endothermic reaction
B. exothermic reaction
C. nuclear fission reaction
D. decomposition reaction

106. Gases with a lighter molecular weight will travel faster than gases with a heavier molecular weight. A container with a mixture of hydrogen gas, helium gas, and oxygen gas in equal amounts has a hole poked in it, allowing the gases to escape. Which of the following is true?

A. After 1 minute, more oxygen will have escaped than helium or hydrogen.
B. After 30 seconds, more helium will have escaped than oxygen or hydrogen.
C. After 3 minutes, there will be less hydrogen in the container than either oxygen or helium.
D. All three gases will escape from the container in equal amounts.

Questions 107–109 are based on the following concept.

Substances can be identified by their properties. There are two basic types of properties that we associate with matter. These properties are called **physical** properties and **chemical** properties. Physical properties are those that you can observe without changing the identity of the substance. Chemical properties are those that can be observed only during a chemical reaction that alters the substance's chemical identity and produces one or more new substances.

107. Indicate the box in which each of the following items belongs.

- Density
- Color
- Flammability
- Reactivity
- Viscosity
- Acidity

A: Physical properties	B: Chemical properties

*The following question contains two blanks, each marked "Select . . . ▼." Beneath each one is a set of choices. Indicate the choice that is correct and belongs in the blank. (**Note:** On the real GED test, the choices will appear as a "drop-down" menu. When you click on a choice, it will appear in the blank.)*

108. The fact that iron is denser than aluminum is an example of a

Select . . . ▼
chemical
physical

property of iron, and the fact that iron exposed to atmospheric oxygen forms rust is an example of a

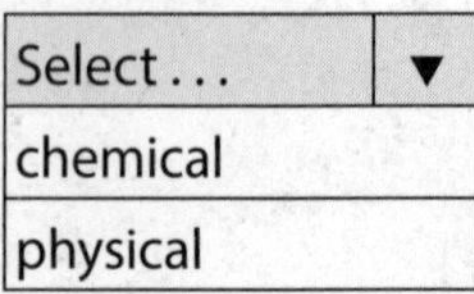

property of iron.

109. Which of the following demonstrates a change in physical properties?

   A. wood burning in a campfire
   B. a banana turning brown
   C. a steak cooking in a frying pan
   D. a glass being shattered by a flying dart

Questions 110–112 are based on the following concept.

Temperature affects the state of a substance. Specific heat causes molecules to vibrate and then move around each other more and more.

110. A scientist places 10 mL of water in a test tube and heats it until the liquid boils and escapes as steam. This is an example of which of the following?

   A. chemical change involving changes of state
   B. chemical change involving chemical reactions
   C. physical change involving changes of state
   D. physical change involving chemical reactions

111. Which of the following is an example of condensation?

   A. the heating of iron until it is liquid
   B. the formation of dew
   C. the making of instant coffee
   D. the making of ice

112. Which of the following statements is true?

   A. Condensation absorbs heat and thus warms the atmosphere.
   B. Evaporation absorbs heat and thus warms the atmosphere.
   C. Condensation releases heat and thus cools the atmosphere.
   D. Evaporation absorbs heat and thus cools the atmosphere.

Questions 113–116 refer to the following passage.

The density of a substance is its mass per unit volume. The symbol for density is the Greek letter ρ (rho).

$$\rho = \frac{m}{V},$$

where ρ is the density, m is the mass, and V is the volume.

Density is a physical property of a substance. For example, a Styrofoam cup is less dense than a ceramic cup because in a ceramic cup, molecules are packed tighter together than in the Styrofoam cup, giving it a higher density. Mixtures can be separated based on the densities of their components. For example, oil rises to the top of salad dressing because it is less dense than the other ingredients.

113. At room temperature, which of the following liquids has the highest density?

A. orange juice
B. milk
C. honey
D. water

114. What is the density of an object with a volume of 15 mL and a mass of 42 g?

A. 0.352 g/mL
B. 2.80 g/mL
C. 630 g/mL
D. 6.30 g/mL

115. Four liquids have densities as follows: Liquid W: 1.5 g/mL; Liquid X: 12 g/mL; Liquid Y: 0.8 g/mL; Liquid Z: 2.8 g/mL. When poured into a container, what is the correct arrangement of these liquids from the bottom of the container to the top?

A. Y, W, Z, X
B. X, Z, W, Y
C. X, W, Z, Y
D. X, Z, Y, W

Write your answer in the blank.

116. Density is a ________________ quantity.

A. variable
B. fundamental
C. scalar
D. vector

Questions 117–120 refer to the following passage.

A chemical equation is a representation of a chemical reaction, where reactants under certain chemical conditions react with each other to form products. A chemical equation is written a specific way to show reactants and products, with an arrow separating reactants from products. In reactions with more than one reactant or product, plus signs separate the individual products and reactants from each other. Numbers called coefficients, placed in front of each reactant, specify the ratio required to convert all of the reactants to products. Products also have coefficients that indicate the ratio of each of the products in the reaction.

The subscripts after a chemical element tell how many atoms of that element are present in the molecule. For example, the chemical equation for the formation of sugar from water and carbon dioxide is as follows:

$$6CO_2 + 6H_2O \rightarrow C_6H_{12}O_6 + 6O_2$$

The equation obeys the law of conservation of mass and is called a balanced chemical equation. The law states that the mass of the reactants must equal the mass of the products. This can be determined by counting the number of atoms of each element to verify that there is the same number of atoms in the reactants as in the products.

117. Ozone (O_3) is produced in the upper atmosphere when oxygen gas (O_2) absorbs high-energy sunlight. Which equation correctly describes this reaction?

A. $O_2 + \text{energy} \rightarrow O_2$
B. $O_2 + \text{energy} \rightarrow 2O_2$
C. $2O_2 + \text{energy} \rightarrow 3O_2$
D. $3O_2 + \text{energy} \rightarrow 2O_3$

118. Iron (Fe) can combine with oxygen (O) to produce any one of the three different iron oxides: FeO, Fe_2O_3, and Fe_3O_4. Of the three oxides, FeO is the most unstable. If FeO is exposed to air, it forms Fe_2O_3. Which equation correctly describes this reaction?

A. $FeO + O_2 \rightarrow Fe_2O_3$
B. $FeO + O_2 \rightarrow 2Fe_2O_3$
C. $2FeO + O_2 \rightarrow Fe_2O_3$
D. $4FeO + O_2 \rightarrow 2Fe_2O_3$

*The following question contains two blanks, each marked "Select . . . ▼." Beneath each one is a set of choices. Indicate the choice that is correct and belongs in the blank. (**Note:** On the real GED test, the choices will appear as a "drop-down" menu. When you click on a choice, it will appear in the blank.)*

119. In a chemical equation, the symbol representing the term *reacts with* is a(n) Select . . . ▼, and the symbol representing the word *yields* is

Select . . . ▼
equals sign
plus sign
coefficient
arrow

a(n) 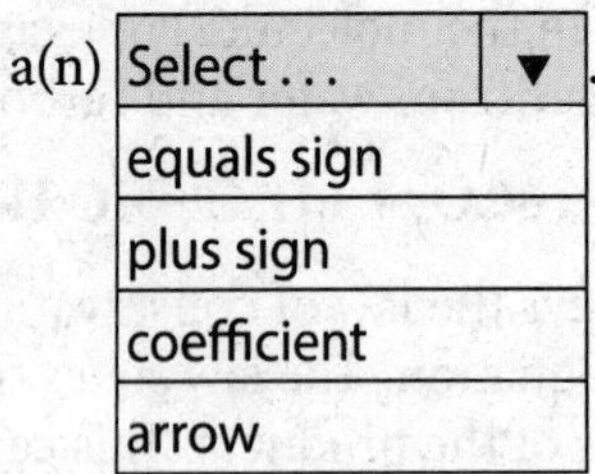

.

Write your answer in the blank.

120. In a chemical equation, the number of molecules of a given substance is indicated by a ____________.

Questions 121–123 refer to the following chart.

Reaction Type	Explanation	Examples
Combination	Two or more compounds combine to form one compound.	Iron and sulfur combine to form iron(II) sulfide: $8Fe + S_8 \rightarrow 8FeS$
Decomposition	A molecule breaks down into simpler ones.	Electrolysis of water to make oxygen and hydrogen gas: $2H_2O \rightarrow 2H_2 + O_2$
Neutralization	An acid reacts with a base. Generally, the product of this reaction is a salt and water.	Hydrobromic acid (HBr) reacts with sodium hydroxide: $HBr + NaOH \rightarrow NaBr + H_2O$
Combustion	A substance reacts with oxygen, releasing energy as light and heat.	Burning of napthalene: $C_{10}H_8 + 12O_2 \rightarrow 10CO_2 + 4H_2O$
Displacement	One element trades places with another element in a compound.	Magnesium replaces hydrogen in water to make magnesium hydroxide and hydrogen gas: $Mg + 2H_2O \rightarrow Mg(OH)_2 + H_2$

121. The following chemical equation takes place in the presence of specific heat.

$CaCO_3(s) \rightarrow CaO(s) + CO_2(g)$

This reaction is an example of a

A. redox reaction
B. combination reaction
C. decomposition reaction
D. displacement reaction

Write your answers in the blanks.

122. In a neutralization reaction, acids react with ________________ to form ________________ and water.

123. Which of the following reactions is a combination reaction?

A. $AgNO_3(aq) + HCl(aq) \rightarrow AgCl(s) + HNO_3(aq)$
B. $Na_2O(s) + CO_2(g) \rightarrow Na_2CO_3(s)$
C. $C_3H_8(g) + 5O_2(g) \rightarrow 3CO_2(g) + 4H_2O(l)$
D. $2H_2O(l) \rightarrow 2H_2(g) + O_2(g)$

Questions 124–126 refer to the following passage.

In a solution, two or more substances are joined in a homogeneous mixture with generally uniform physical properties. In such a mixture, a *solute* is a substance dissolved in another substance, known as a *solvent*. The solvent is the most plentiful substance in a solution. For example, when sugar is dissolved in water, sugar is the solute and water is the solvent.

The ability of one substance to dissolve in another substance is called its solubility. When a given amount of solvent contains as much solute as it can hold, the solution is said to be saturated.

Various factors affect solubility, including temperature and pressure. Solubility increases with rising temperature in solids, whereas gases are usually more soluble at colder temperatures. For solids and liquids, the pressure dependence of solubility is typically weak, whereas for gases, solubility increases with pressure.

*The following two questions contain a blank marked "Select . . . ▼." Beneath it is a set of choices. Indicate the choice that is correct and belongs in the blank. (**Note:** On the real GED test, the choices will appear as a "drop-down" menu. When you click on a choice, it will appear in the blank.)*

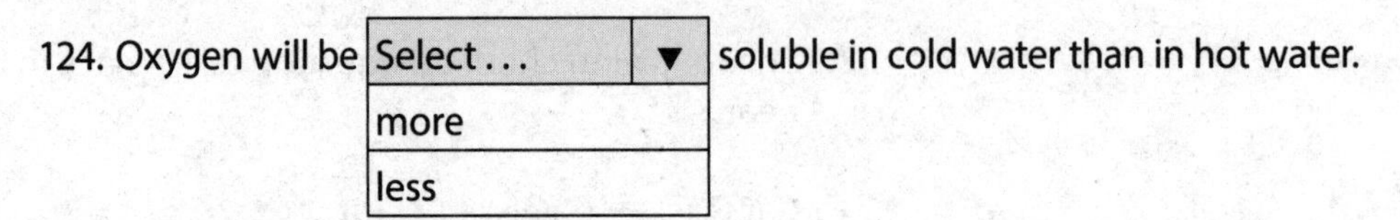

125. Atmospheric air is a homogeneous mixture of gases that is mostly nitrogen gas. Therefore, nitrogen in the atmosphere can be considered the.

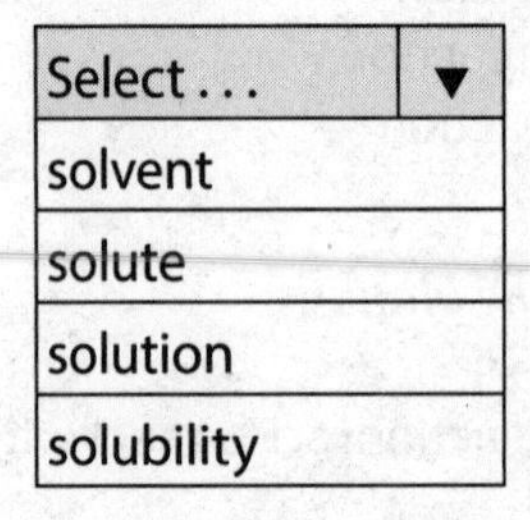

126. Which of the following would increase the solubility of nitrogen?

A. increasing the temperature
B. mixing it with an equal amount of oxygen
C. decreasing pressure
D. increasing pressure

127. What temperature in degrees Fahrenheit is it if the thermometer reads 25 degrees Celsius? Use the formula $F = \frac{9}{5}C + 32$.
   A. 32°
   B. 57°
   C. 65°
   D. 77°

128. Which of the following describes isotopes?
   A. Two atoms have the same number of neutrons and the same number of protons.
   B. Two atoms have different numbers of neutrons, but the same number of protons.
   C. Two atoms have the same number of neutrons, but different numbers of protons.
   D. Two atoms have the same number of neutrons, but different numbers of electrons.

129. Which of the following describes the process of diffusion?
   A. A substance moves from an area of low concentration to an area of high concentration.
   B. A substance moves from an area of high concentration to areas of low concentration.
   C. A substance moves from the center toward two opposite poles.
   D. A liquid is turned into a gas through the application of heat.

130. Which of the following substances has a neutral pH?
   A. lemon juice with a pH of 2
   B. water with a pH of 7
   C. soapy water with a pH of 12
   D. milk with a pH of 6

131. Which of the following describes an ionic bond?
   A. Two atoms of the same element share a pair of electrons in the middle of the two atoms.
   B. One metallic and one nonmetallic atom bond by attracting oppositely charged ions.
   C. Two metallic atoms bond through the electrostatic attractive force between conduction electrons and positively charged metal ions.
   D. Two nonmetallic atoms bond by sharing their electrons.

Use the following information to answer Questions 132–133.

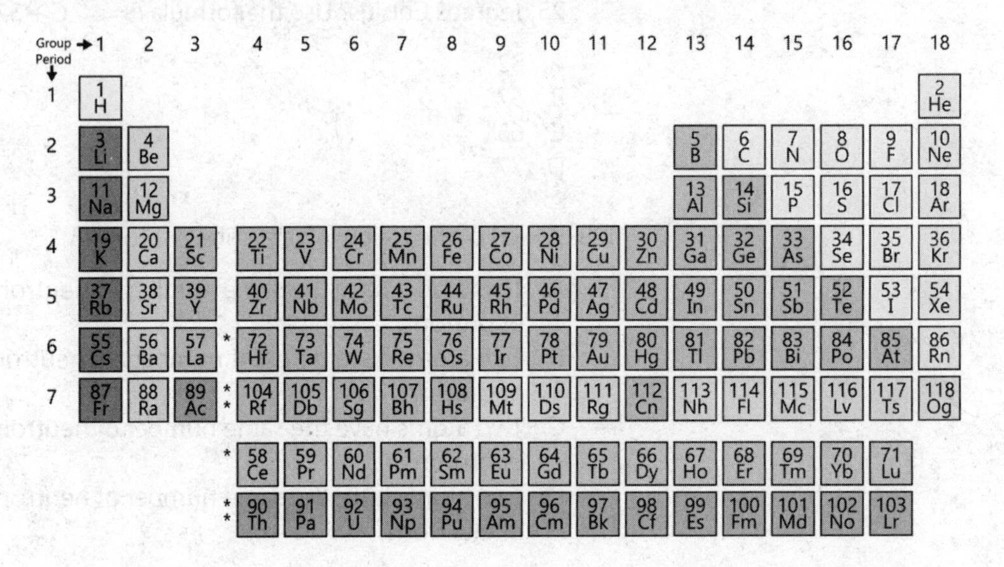

132. Which two elements are expected to have similar chemical properties?

A. C and B
B. H and He
C. Ca and Sr
D. N and Ne

133. Which of the following atoms is the most stable?

A. nitrogen, because the elements in Group 15 can form anions
B. argon, because the noble gases have full valence shells
C. sodium, because alkali metals can form cations
D. magnesium, because alkaline earth metals have a high melting point

134. Which equation shows the synthesis of water?
Place an X on the blank next to your choice.

_________ $2Na + Cl_2 \rightarrow 2NaCl$
_________ $2H_2 + O_2 \rightarrow 2H_2O$
_________ $NaOH \rightarrow Na + OH$
_________ $4Fe + 3O_2 \rightarrow 2Fe_2O_3$

135. In an experiment to test whether more time spent studying for a test will increase test scores, which of the following is the independent variable?

A. the subject matter of the test the students are given
B. the number of questions on the test
C. the scores the students make on the test
D. the amount of time each student spends studying

136. A blue shift of the light from a star indicates that the star

   A. will soon explode in a supernova.
   B. will become a black hole.
   C. is moving toward Earth.
   D. is moving away from Earth.

Use the following information to answer Question 137.

To start a fire with an artificial flint sparking stick, you will first need some tinder to start your fire. There are many things you can use for tinder, but it must be completely dry. You can use dry grass or leaves, tree bark or small sticks, paper, cloth, and so on. Use your knife to shred the tinder finely so that it will ignite from a spark. Once you have the tinder, put a bit of it into a small pile, and place the sparking stick close to the tinder. Use the edged tool that comes attached to the sparking stick and draw it slowly downward over the sparking stick. If your flint does not have this tool, you can use your knife blade. Dragging the knife blade or edged tool across the surface of the sparking stick will produce hot sparks that will light the tinder if you have prepared it properly. Finally, when the fire is becoming established, you should gradually add more tinder, then thin sticks of wood. Larger pieces of wood should only be added once the fire is going well. Build your fire up gradually.

137. Place the following steps in the order they should be done. (***Note***: *On the real GED science test, you will click on each sentence and "drag" it into position in the chart.*)

Order of Events

1.
2.
3.
4.

   A. Gathering dry leaves, small twigs, and grass
   B. Adding larger pieces of wood
   C. Using the sparking stick to produce sparks
   D. Shredding the tinder with your knife

138. Which of the following reactions needs the most manipulation to make sure that the reactants do *not* re-form from the products?

   A. $HCl + NaOH \rightarrow NaCl + H_2O$
   B. $H_2(g) + I_2(g) \longleftrightarrow 2HI(g)$
   C. $AgNO_3(aq) + NaCl(aq) \rightarrow NaNO_3(aq) + AgCl(s)$
   D. $Zn(s) + 2HCl \rightarrow ZnCl_2(aq) + H_2(g)$

139. A benefit of using nuclear reactions to produce electricity is that

   A. nuclear waste is safe and easy to dispose of.
   B. no harmful greenhouse gases are produced.
   C. any radioactive materials released into the atmosphere by accident are harmless.
   D. the materials used to produce the energy can also be used to make harmful weapons.

140. A solid, liquid, and gas can exist at the same time at the

   A. flash point.
   B. boiling point.
   C. critical point.
   D. triple point.

CHAPTER 3

Earth and Space Science

Chapter 3 of this *Science Workbook* covers topics in Earth and space science. These topics constitute about 20 percent of the science test. Earth science includes the study of rocks, soil, and water and interactions between living and nonliving things on Earth.

The four basic areas of Earth science covered on the GED Science test are geology, meteorology, oceanography, and astronomy. Geology topics covered include the interior structure of Earth, the movement of tectonic plates that produces landforms such as mountains and ocean basins, and Earth's rock, hydrologic, and other systems. Meteorology topics include the characteristics of the atmosphere, the various layers of the atmosphere, climate and weather, and the causes and effects of climate change. Oceanography topics covered include the characteristics of oceans and their effects on Earth and its organisms. Other topics covered in these areas include resource extraction processes; the uses of natural, renewable, and nonrenewable resources; and the causes and effects of natural hazards, including hurricanes and earthquakes.

The other major topic covered on the science test is the cosmos—its structure and organization. Questions focus on major structures in the universe, such as the galaxies, stars, constellations, and solar system. Other topics include the age and development of the universe, stellar movement, and the life cycles and deaths of stars (black holes and white dwarfs).

Directions: Answer the following questions. For multiple-choice questions, choose the best answer. For other questions, follow the directions preceding the question. Answers begin on page 176.

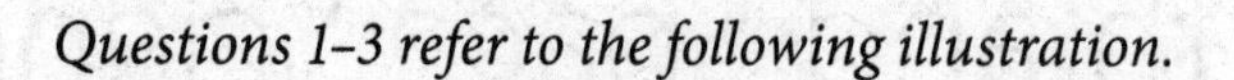

Questions 1–3 refer to the following illustration.

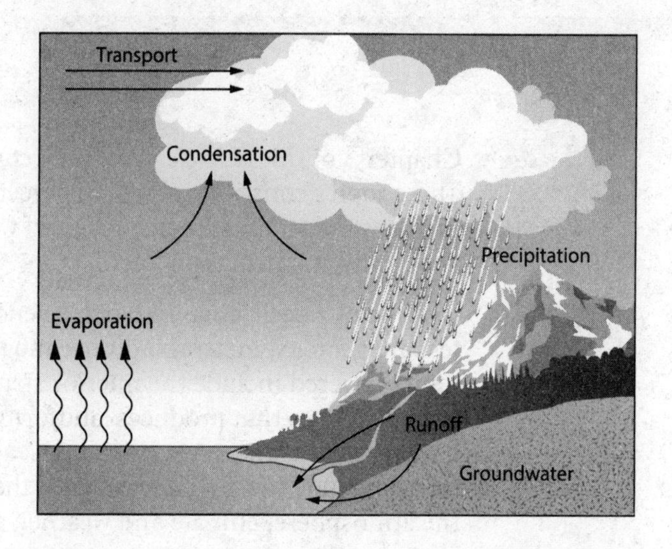

*The following question contains a blank marked "Select . . . ▼." Beneath it is a set of choices. Indicate the choice that is correct and belongs in the blank. (**Note**: On the real GED test, the choices will appear as a "drop-down" menu. When you click on a choice, it will appear in the blank.)*

1. The biogeochemical cycle pictured represents the Select . . . ▼ cycle.

Select . . . ▼
carbon
nitrogen
water
phosphorus

2. What term describes the release of water vapor from the leaves of plants?

   A. evaporation
   B. precipitation
   C. transpiration
   D. condensation

3. Which of the following describes precipitation?

   A. Water rises up into the atmosphere as vapor.
   B. Water vapor in the air forms clouds.
   C. Trees drop their leaves in winter.
   D. Water falls from clouds as rain or snow.

Questions 4 and 5 refer to the following passage.

Carbon is continually cycled within Earth. Carbon is a life-sustaining element and a key component of many systems in the biosphere. Carbon is part of Earth's thermostat and one of the key components of life. Through the carbon cycle, carbon moves between the atmosphere, oceans, plants and animals, and soil, and within the Earth. Disruption of this balance can have serious consequences, such as climate change.

The Carbon Cycle

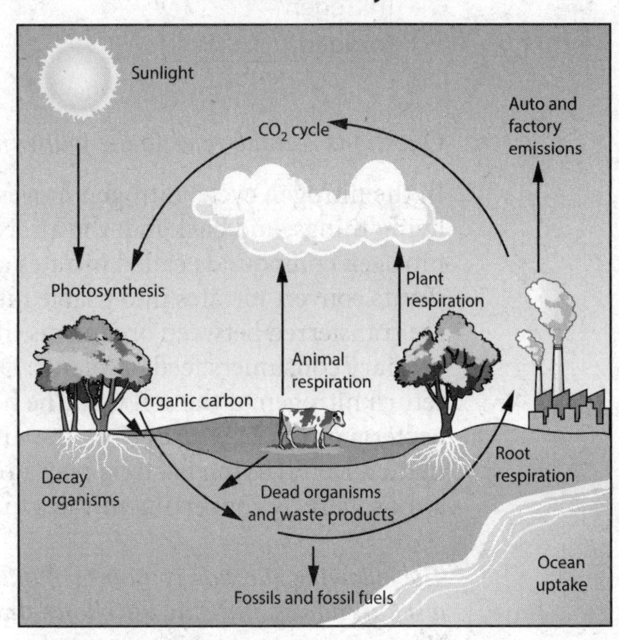

4. In the carbon cycle, fossil fuels are produced from

   A. factory emissions
   B. photosynthesis
   C. animal respiration
   D. the decay of organisms that were once living

5. Which is NOT part of the carbon cycle?

   A. In the ocean uptake process, the oceans absorb carbon from the atmosphere.
   B. Carbon atoms enter the ground as dead organisms and animal waste products.
   C. Plants gather carbon dioxide from the soil where organic carbon is present in the form of manure and fertilizers.
   D. Animals consume plants and thus take in carbon atoms, which the animals release into the atmosphere through respiration.

*The following question contains a blank marked "Select... ▼." Beneath it is a set of choices. Indicate the choice that is correct and belongs in the blank. (**Note**: On the real GED test, the choices will appear as a "drop-down" menu. When you click on a choice, it will appear in the blank.)*

6. Permanent deforestation can contribute to global warming by increasing Select... ▼ levels in the atmosphere.

Select... ▼
water
carbon dioxide
nitrogen
oxygen

Questions 7 and 8 refer to the following passage.

In the nitrogen cycle, nitrogen moves from the air into the soil, into living things, and back into the air. Nitrogen enters the food chain when nitrogen compounds called nitrates are absorbed from the soil by plants. Plants convert nitrates into usable nitrogen compounds. These compounds are transferred between organisms through feeding levels starting when primary consumers feed on plants. Decaying plants and animals also return nitrogen to the soil with the help of nitrogen-fixing bacteria. These bacteria produce nitrates from the nitrogen compounds found in decaying matter. Denitrifying bacteria play a role when they return nitrogen to the atmosphere by converting nitrates to nitrogen gas.

*The following question contains a blank marked "Select... ▼." Beneath it is a set of choices. Indicate the choice that is correct and belongs in the blank. (**Note**: On the real GED test, the choices will appear as a "drop-down" menu. When you click on a choice, it will appear in the blank.)*

7. Nitrogen-fixing bacteria in the soil produce nitrates from the nitrogen in Select... ▼.

Select... ▼
living creatures
denitrifying bacteria
decaying plants

8. Number the following steps in the nitrogen cycle in order from 1 to 5. Step 1 is done for you.

Write the numbers in the spaces provided.

____1____ Bacteria in the soil remove nitrogen from decaying plants and animals.

________ Herbivore animals feed on plants and get essential nitrogen.

________ Plants absorb nitrates from the soil.

________ Plants and animals die and their remains decay.

________ Bacteria produce nitrogen compounds called nitrates.

Questions 9–11 refer to the following passage.

Fossil fuels, which are coal, fuel oil, and natural gas, are hydrocarbons formed from the remains of dead plants and animals.

As trees and plants died and sank to the bottom of swamps and oceans, they formed layers of a spongy material called peat. The peat became covered by sand and clay, which over time became sedimentary rock. As more rock formed, it pressed down on the peat. The increased pressure over millions of years turned the peat into fossil fuels.

9. Which of the following is NOT a fossil fuel?
   A. coal
   B. petroleum
   C. wood
   D. methane

10. Coal is formed when
    A. minerals replace the organic matter in dead plants and animals, creating fossils
    B. deposits of methane are subjected to high pressure until they condense into crystals
    C. swamp plants are buried by sediment, producing heat and pressure that convert the organic matter to coal
    D. organic matter from dead plants and animals becomes trapped within layers of sedimentary rock

*The following question contains a blank marked "Select . . . ▼." Beneath it is a set of choices. Indicate the choice that is correct and belongs in the blank. (**Note**: On the real GED test, the choices will appear as a "drop-down" menu. When you click on a choice, it will appear in the blank.)*

11. Fossil fuels are formed within Earth's crust by natural processes. However, they are considered nonrenewable because

Select . . . ▼
they are extremely rare
they take a very long time to form
they are very difficult to extract

Questions 12–14 refer to the following information.

The burning of fossil fuels has increased the amount of carbon dioxide, a greenhouse gas, in the atmosphere. This gas has caused average temperatures on Earth to rise because it traps heat that would normally escape back into space. The levels of carbon dioxide have been estimated at the following concentrations in parts per million over recent years:

Year	1960	1970	1980	1990	2000	2010	2013
CO_2 Level	315	325	340	350	370	390	400

In addition to the threat of rising global temperatures, carbon dioxide can also produce acid rain as raindrops mix with the carbon dioxide and precipitate to Earth's surface. The equation for this reaction is

$$CO_2(g) + H_2O(l) \rightarrow H_2CO_3(aq)$$

12. Which of the following is a way to combat the increase in levels of carbon dioxide in the atmosphere?
    A. using alternative energies that are renewable
    B. drilling for offshore oil
    C. cutting down trees in the Amazon forest
    D. shipping oil via pipelines instead of via oil tankers

13. When carbon dioxide reacts with rain water, the end product will most likely
    A. have no impact on the environment over time
    B. raise the acidity level of lakes and streams, producing a harmful impact on wildlife
    C. be a basic liquid that is harmless to the environment
    D. help in preserving marble statues

14. A scientist observes that the fish population in a local lake has drastically decreased in recent years. On testing the water, it is found to be unusually acidic. The scientist hypothesizes that acid rain is the culprit, but there are no obvious sources nearby. What data might the scientist collect to support his or her hypothesis?
    A. data about industries that are considering relocation to the area
    B. data about fish predators
    C. data about the amount of rainfall over the past few years
    D. data about weather patterns

15. Which of the following was most likely formed via the action and movement of a glacier?
    A. V-shaped valley
    B. volcano
    C. U-shaped valley
    D. plateau

16. Water can cause weathering and erosion. The products most likely to form are

 A. igneous rocks
 B. metamorphic rocks
 C. sedimentary rocks
 D. fine sediments

Questions 17 and 18 refer to the following passage and the diagram.

A fault is a disruption in the Earth's crust along which a movement can take place, causing an earthquake. A sudden slip of a fault, volcanic activity, or other sudden stress or changes in the Earth can cause an earthquake.

There are various terms related to earthquakes and faults. The **focus** is the point within Earth where the earthquake rupture starts. The **epicenter** is the point on Earth's surface directly above the point where a seismic rupture begins. **Dip** is the angle that the fault is inclined from the horizontal. The **fault scarp** is the feature on the surface of the Earth that looks like a step caused by a slip in the fault.

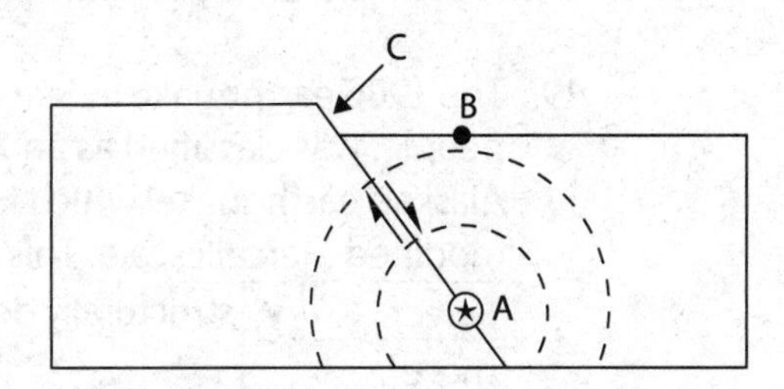

17. In the diagram, where is the focus with respect to the epicenter?

 A. directly below the epicenter
 B. directly above the epicenter
 C. between the epicenter and the dip
 D. between the epicenter and the fault scarp

18. Identify points A, B, and C in the diagram.

 A. Point A represents the focus. Point B represents the epicenter. Point C represents the fault scarp.
 B. Point A represents the epicenter. Point B represents the focus. Point C represents the fault scarp.
 C. Point A represents the epicenter. Point B represents the focus. Point C represents the dip.
 D. Point A represents the fault. Point B represents the epicenter. Point C represents the focus.

Questions 19 and 20 refer to the following passage.

Seismic waves are the vibrations from earthquakes that travel through Earth; they are recorded on instruments called seismographs. Seismographs record a zigzag trace that shows the varying amplitude of ground oscillations beneath the instrument. The Richter scale is a scale that is used to measure the magnitude of an earthquake as determined by seismographic readings.

The modified Mercalli scale is used to measure the intensity of an earthquake. The modified Mercalli scale uses some key responses to the earthquake, for example, people waking up, furniture movement, and the extent of damage to structures. It does not have a mathematical basis. It is an arbitrary ranking based on observed effects. The lower numbers of the intensity scale relate to how the earthquake is felt by people. The higher numbers of the scale are based on observed damage to buildings.

*The following question contains a blank marked "Select . . . ▼." Beneath it is a set of choices. Indicate the choice that is correct and belongs in the blank. (**Note**: On the real GED test, the choices will appear as a "drop-down" menu. When you click on a choice, it will appear in the blank.)*

19. The 1906 earthquake in San Francisco, California, which killed 3,425 people, was classified as an X on the modified Mercalli scale. The 1964 Alaskan earthquake, which killed 143 people, was classified as an XI on the modified Mercalli scale. This means that the Alaskan earthquake was Select . . . ▼ structurally destructive than was the San Francisco earthquake.

Select . . . ▼
more
less

Write your answers in the blanks.

20. The magnitude of an earthquake can be expressed by the ________________ scale, and the intensity of an earthquake can be measured by the ________________ scale.

Questions 21–23 refer to the following passage.

Intense storms can develop over warm tropical oceans. The scientific name for these storms is tropical cyclones, but they are also referred to as hurricanes in North America and typhoons in Asia.

Tropical cyclones comprise warm, moist air formed only over warm ocean waters near the equator. The warm, moist air over the ocean rises upward from the surface, causing an area of lower air pressure below. Wind with higher pressure rushes toward the center. This continues and as the warm air continues to rise, the surrounding air swirls in to take its place. During this entire process of warm, moist air rising and cooling, the water in the air forms clouds and spiral rain bands. The whole system of clouds and wind spins and grows, fueled by the ocean's heat and evaporating water. As the storm rotates faster and faster, an eye is formed in the center. The center of the storm, or eye, is the calmest part. It has only light winds and fair weather. The border of the eye is the eye wall, where the most damaging winds and intense rainfall are found. Cyclones in the Northern Hemisphere move counterclockwise, while those in the Southern Hemisphere move clockwise.

21. What is the atmospheric pressure at the eye of a tropical cyclone?

    A. very high because of converging winds
    B. very high because of sinking warm air
    C. very low because of rising warm air
    D. very low because of dense sinking air

22. Choose the correct label from the list for features A, B, C, and D shown in the diagram.

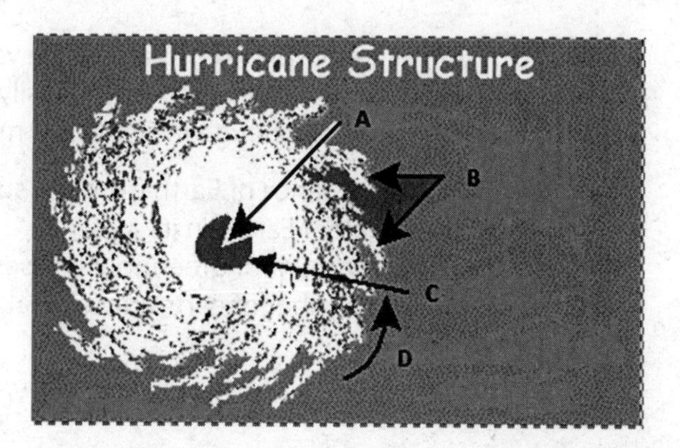

eye   eye wall   clockwise rotation   counterclockwise rotation   spiral rain bands   clouds

Write your answers on the lines.

A. ______________________

B. ______________________

C. ______________________

D. ______________________

*The following question contains two blanks, each marked "Select . . . ▼." Beneath each one is a set of choices. Indicate the choice that is correct and belongs in the blank. (**Note**: On the real GED test, the choices will appear as a "drop-down" menu. When you click on a choice, it will appear in the blank.)*

23. A tropical storm that develops over the Select . . . ▼ Ocean can

Select . . . ▼
Atlantic
Pacific
Indian
Antarctic

cause a hurricane in the eastern United States, whereas a storm that develops over the Select . . . ▼ Ocean can

Select . . . ▼
Atlantic
Pacific
Indian
Antarctic

cause a typhoon in the Philippines and Hong Kong.

24. Why are dikes and dams built along the hurricane-prone coastline of the United States?
    A. to change the course of rivers that flow into the Atlantic Ocean
    B. to keep people from swimming in the ocean during hurricanes
    C. to keep land from flooding due to storms
    D. to generate electricity

25. Different places on Earth have various seasons throughout the year. The change in these seasons is easily predictable based upon the date of the year. The change in seasons is most dependent upon the

    A. distance of Earth from the sun
    B. tilt of Earth on its axis
    C. amount of greenhouse gases in the atmosphere
    D. number of earthquakes that have occurred over 3 months

26. The population of humans on Earth has grown exponentially over time. It is believed that eventually the population will reach the maximum number that Earth's resources can sustain. This number is called Earth's

    A. carrying capacity
    B. extinction
    C. thriving population
    D. homeostasis

27. Early in Earth's history, it is believed that Earth's atmosphere contained a number of gases. These gases combined to form the first large organic compounds. This is referred to as the "hot thin soup" concept and is illustrated in the following diagram.

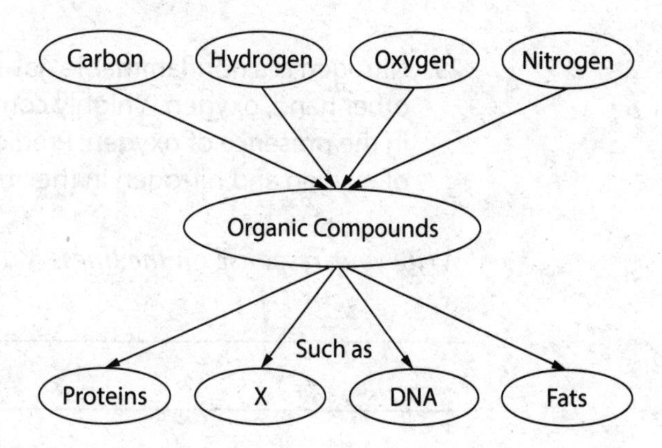

    In the diagram, substance X is most likely

    A. carbon dioxide
    B. ozone gas
    C. glucose
    D. ammonia

28. A red shift of the light from a star indicates that the star

    A. will soon explode in a supernova
    B. will become a black hole
    C. is moving toward Earth
    D. is moving away from Earth

Refer to the following circle graph to answer Question 29.

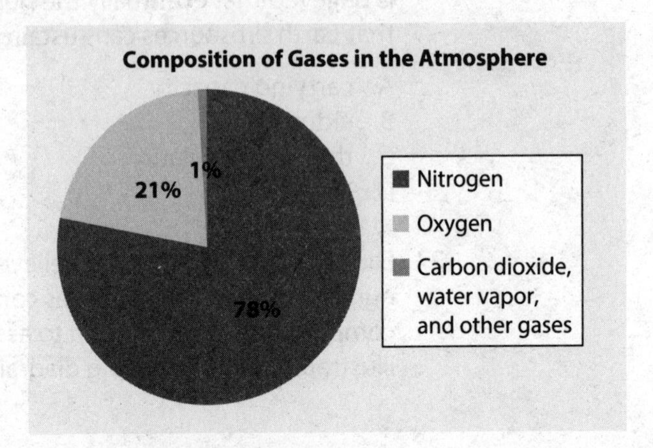

29. Nitrogen is a nonflammable gas and does not support combustion. On the other hand, oxygen is highly combustible. All flammable substances burn in the presence of oxygen. Predict what could happen if the percentages of oxygen and nitrogen in the atmosphere were reversed.

Write your response on the lines.

____

____

____

____

____

____

____

____

____

____

____

Answer Questions 30 and 31 based on the following information.

The atmosphere has four major layers.

- **Troposphere**—This is the lowest layer of the atmosphere, extending up to about 6 miles. This is the only layer where water vapor is present, and thus all kinds of weather phenomena occur here. In this layer, temperature decreases at the rate of 1°C for every 165 meters of ascent.
- **Stratosphere**—Above the troposphere lies the stratosphere, which extends up to about 31 miles. Ozone is present in this layer, which helps in the absorption of ultraviolet radiation.
- **Mesosphere**—This is the third layer of the atmosphere and extends up to a height of 50 miles. Most meteorites burn here because of friction.
- **Thermosphere**—This layer extends to a height of about 435 miles. Gas molecules exhibit high kinetic energy, and thus the temperature is high in this layer. Within the lower layers of the thermosphere is the ionosphere.

30. Why does atmospheric temperature increase with altitude in the stratosphere?

    A. because it is closer to the sun
    B. because is it less dense than the troposphere
    C. because it is under a lot of pressure
    D. because of the presence of the ozone layer

31. A plane flies at a height of 37,000 feet to avoid any hindrance due to weather conditions. In which layer of the atmosphere does the plane travel?

    A. thermosphere
    B. mesosphere
    C. stratosphere
    D. troposphere

Refer to the following information to answer Questions 32 and 33.

The following table shows three temperature readings taken on Sierra Peak, a 10,540-foot-high mountain.

Altitude (in feet)	Temperature (in °C)
3,000	20.1
4,000	15.5
5,000	10.3

*The following question contains a blank marked "Select . . . ▼." Beneath it is a set of choices. Indicate the choice that is correct and belongs in the blank. (**Note:** On the real GED test, the choices will appear as a "drop-down" menu. When you click on a choice, it will appear in the blank.)*

32. The average rate of change of temperature with increase in altitude is Select . . . ▼ for every thousand-foot rise in height.

Select . . . ▼
3°C
4°C
5°C

Write your answer in the blank.

33. Assuming that the identified rate is constant, at about ________________ feet, the temperature will be at the freezing point of water.

Refer to the below information to answer Question 34.

Exposure to ultraviolet rays can cause extreme damage to organisms, including death. However, the presence of ozone in the stratosphere prevents ultraviolet rays from reaching Earth. Oxygen gets converted into ozone in the presence of ultraviolet rays and then ozone absorbs large amounts of these rays. However, a depletion in the ozone has been noted due to increasing pollution, especially due to the CFCs released from refrigerators and air conditioners. This depletion in ozone is allowing more ultraviolet rays to reach Earth, resulting in increased occurrences of cataracts and skin cancer.

34. Which of these can NOT be concluded from the information above?

    A. Ultraviolet rays can kill organisms.
    B. Ozone absorbs ultraviolet rays.
    C. Chemicals such as CFCs have led to the depletion of ozone in the stratosphere.
    D. Ultraviolet rays are the only reason for skin cancer.

Refer to the following information to answer Question 35.

The greenhouse effect results when gases in the atmosphere function as an insulating barrier and absorb radiation that would otherwise be reflected back to the upper atmosphere. In other words, like glass in a greenhouse, the gases admit incoming solar radiation but retard its going out into space. This causes gradual warming of the Earth's surface as well as the lower atmosphere. The greenhouse effect makes it possible for life to exist on Earth.

Earth's climate has changed dramatically over time. But in addition to natural variations, human activity is now rapidly altering the composition of the atmosphere. In particular, the use of fossil fuels for heating, transportation, and industry is producing increasing amounts of greenhouse gases. The result has been a rise in average global temperatures, a process known as global warming. This process, if not controlled, stabilized, or even reversed, is thought likely to produce rapid and profound change in Earth's climate.

35. Which of the following is most likely to reduce the global warming trend?

    A. increasing human population
    B. increasing number of trees
    C. increasing use of fuels
    D. increasing insect population

36. In the hydrologic cycle shown below, match the letters with the correct processes. Write the letters on the blanks.

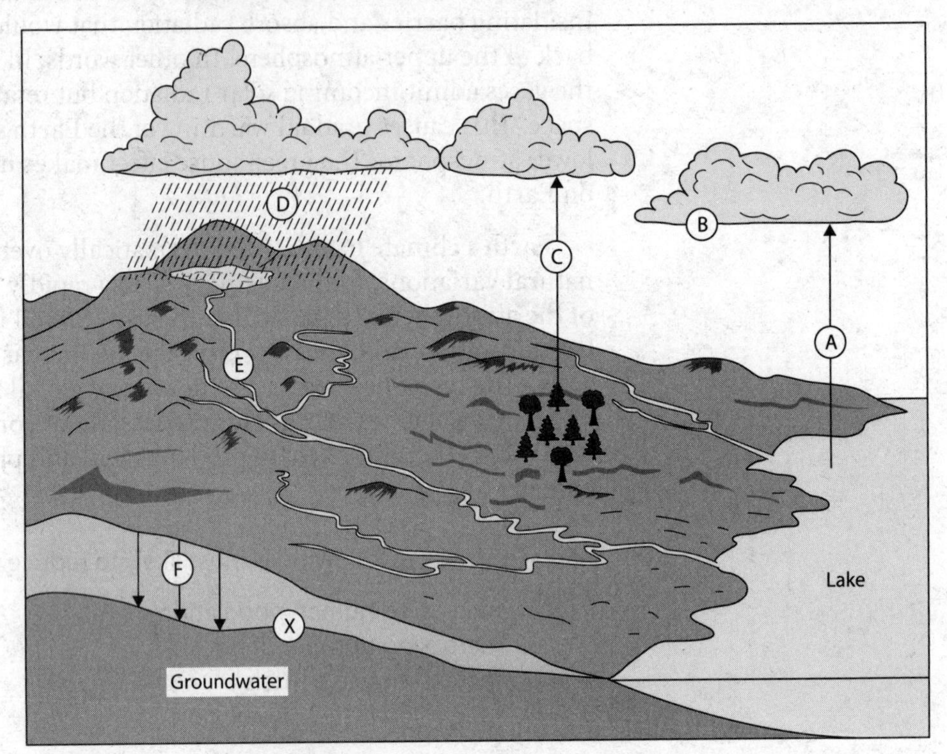

A _______ precipitation
B _______ evaporation
C _______ infiltration
D _______ runoff
E _______ condensation
F _______ transpiration

37. In the past it was believed that our sun and other stars were balls of fire. Today, however, scientists have confirmed that the sun is not fueled by combustion. Instead, the sun is known to be fueled by a nuclear reaction. The nuclear reaction that takes place within the sun is best described as

A. a series of reactions in which chemical bonds are broken to release energy
B. energy production via convection that takes place on the sun's surface
C. the burning of fossil fuels within the sun
D. a series of fusion reactions that take place within the sun

38. Which of the following events is cyclic and highly predictable?

A. a volcano eruption
B. an earthquake
C. Jupiter's movement across the night sky
D. an asteroid striking Earth

Question 39 is based on the following information.

Weathering refers to the breaking down of rocks. This may be due to physical factors such as wearing away of the rock surface by wind and many other natural agents. Chemical processes such as repetitive freezing and liquefaction may lead to weathering of rocks too. Organic factors, that is, living things, may also aid in weathering.

39. Indicate the box in which each of the following agents of erosion belongs.

- Acid rain
- Earthworm
- Rain
- Ocean waves
- Carbonic acid
- Lichens
- Oxygen
- Frost
- Moss

A: Physical	B: Chemical	C: Organic

40. In deserts, one can sometimes see rocks shaped almost like mushrooms. That is, a rock might have a slender cylindrical base topped by a large, bulky mass that looks like the cap sitting on a mushroom stalk. What is the most likely cause of the shape of the mushroom-shaped rocks?

    A. erosion from the impact of wind-blown sand
    B. erosion from the action of rain water flowing through the sand
    C. magma flowing through the sand
    D. erosion from the movement of nearby sand due to earthquake

Questions 41 and 42 are based on the following:

There is evidence that the continents have moved over time. It is believed that at one point in Earth's history, all of the current continents were one landmass called Pangaea.

Based upon current models of the movement of the continents, it is estimated that if Christopher Columbus were to sail today to the Americas as he did in 1492, his ship would have to sail approximately 155 feet farther to reach land.

41. This movement of the continents is best related to

    A. weathering at Earth's surface
    B. the diversity of life on Earth
    C. the increase in the number of humans present on Earth over time
    D. plate tectonics and convection within Earth

42. Which of the following is associated with the movement of two continents away from each other?

    A. the formation of hurricanes
    B. the formation of an ocean ridge
    C. the formation of fjords
    D. the formation of a V-shaped valley

Refer to the following graph to answer Question 43.

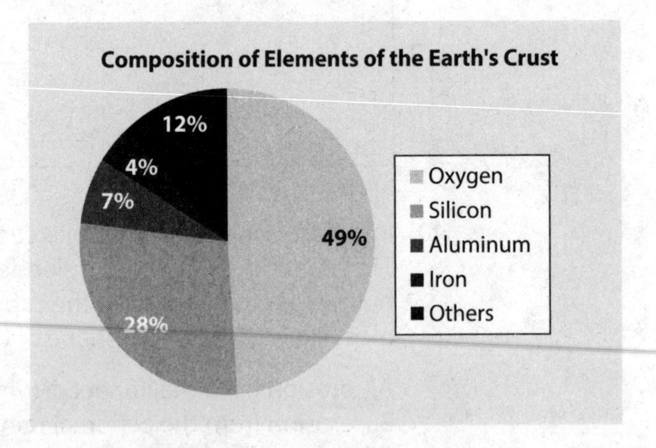

43. Based on the graph, which of the following is true?

    A. The amount of silicon is half that of oxygen.
    B. The amount of silicon is four times that of iron.
    C. The amount of aluminum is one-fourth that of silicon.
    D. The amount of oxygen is four times that of aluminum.

Refer to the following information to answer Question 44.

Constructive plate boundaries represent zones of divergence where there is continuous upwelling of lava and new oceanic crust is continuously formed.

Destructive plate boundaries converge along a line and the leading edge of one plate overrides the other. This overridden plate gets subducted into the mantle, forming deep trenches.

Conservative plate boundaries are formed when two plates slide past one another along transform faults and thus crust is neither created nor destroyed.

Write your answer in the blank.

44. The Mariana Trench is located to the west of the Philippines in the North Pacific Ocean. It is the boundary between an oceanic plate and a continental plate. This crust formation is the result of a
_________________ boundary.

*The following question contains two blanks, each marked "Select . . . ▼." Beneath each one is a set of choices. Indicate the choice that is correct and belongs in the blank. (**Note**: On the real GED test, the choices will appear as a "drop-down" menu. When you click on a choice, it will appear in the blank.)*

45. Molten rock above Earth's surface is called Select . . . ▼,

Select . . . ▼
lava
magma

while molten rock below Earth's surface is called Select . . . ▼.

Select . . . ▼
lava
magma

Refer to the following information to answer Questions 46–48.

The Mid-Atlantic Ridge is a long underwater mountain ridge that sits about halfway between the continents on either side of it. Scientists believe that the sea floor is spreading outward along this ridge. This spreading seems to be caused by the continual flow of magma from the cracks in Earth's crust and from eruptions of volcanoes along the mountain chain. Because of the sea floor spreading, North and South America are slowly moving farther from Europe and Africa.

46. What is the probable cause of spreading of the sea floor at the Mid-Atlantic Ridge?

    A. the movement of continents, followed by underwater earthquakes
    B. the movement of tectonic plates with an accompanying outpouring of magma
    C. the eruption of underwater volcanoes and an outpouring of lava
    D. the movement of ocean currents due to the rotation of Earth and its axis

47. The map below shows the tectonic plate boundaries near the Rift Valley of eastern Africa. The arrows show the movement of the plates. A region of Africa is crosshatched (⨳). What is happening to the crosshatched region of eastern Africa?

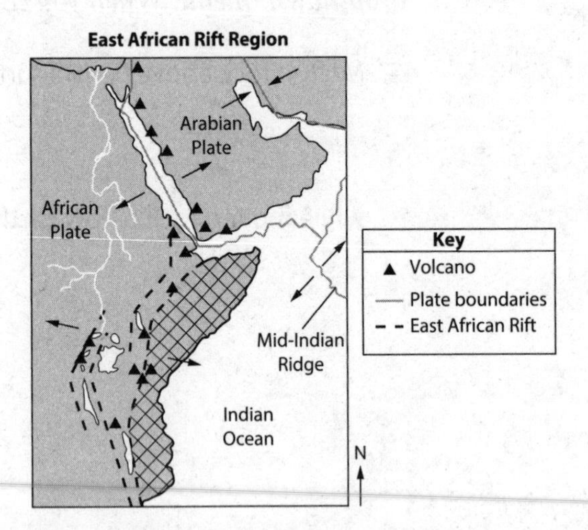

    A. This region is colliding with the rest of Africa, forming mountains.
    B. This region is colliding with the Arabian Plate.
    C. This region is moving eastward relative to the rest of Africa.
    D. This region is moving northward relative to the rest of Africa.

48. What is the best evidence that sea floor spreading occurs at about the same rate on both sides of the ridge?

    A. The shape of South America is similar to the shape of Africa.
    B. The continents to the left of the ridge are about the same distance from the ridge as the continents to the right.
    C. The sea floor to the left of the ridge contains the same minerals as the sea floor to the right.
    D. The ridge is the same general shape as the shorelines on each side of it.

Refer to the following information to answer Questions 49 and 50.

The universe is now thought to have originated at a time billions of years ago when all the matter and energy that exists was concentrated together at one single tiny point. The matter was extremely hot, and it burst outward into space in a huge explosion (the "Big Bang"). As the matter flowed outward, over time it coalesced into simple atoms, elements, and finally stars and galaxies.

What is the evidence for the Big Bang? Spectral lines provide the key. Light from objects that are moving away from Earth produces spectral lines that are shifted slightly toward the red side of the spectrum. Light from most galaxies shows this so-called red shift, meaning that the galaxies are continuously moving farther away. Scientists think that their movement is a result of the Big Bang. Galaxies are moving farther and farther apart from each other, and the universe is constantly expanding.

Write your answer in the blank.

49. The evidence that the universe is expanding can be found in the ________________ shift of light spectra from distant galaxies.

*The following question contains a blank marked "Select... ▼." Beneath it is a set of choices. Indicate the choice that is correct and belongs in the blank. (**Note**: On the real GED test, the choices will appear as a "drop-down" menu. When you click on a choice, it will appear in the blank.)*

50. Scientists believe that all of the mass and energy of the universe was once concentrated together at one single tiny point that exploded. This is the Select... ▼ theory.

Select... ▼
Red Shift
Hot, Thin Soup
Big Bang
Supernova

Refer to the following information to answer Questions 51 and 52.

In planetary systems like our solar system, the planets orbit around a massive star at the center. Each planet travels at its own speed and at its own distance from the central star. The German astronomer Johannes Kepler determined in the early 1600s that the planets in our solar system move in slightly elliptical orbits. The orbits are not perfectly circular. Kepler also determined that there is a relationship between a planet's distance from the sun and the speed at which it travels. If mass is equal, an object closer to the star will travel faster than an object farther away.

The English physicist Isaac Newton determined that it is the force of gravity that keeps the planets in orbit around the sun and moons in orbit around the planets. Every object with mass exerts a gravitational force of attraction on other objects. Earth and its moon exert gravitational forces on each other. But Earth and the moon do not crash into each other. Instead, the moon stays in orbit because of its constant forward motion. The moon's motion counteracts the force of Earth's gravity. The moon comes no closer to Earth, but at the same time Earth's gravity keeps the moon revolving around Earth rather than traveling off into space.

51. How do the sun and the moon cause tides?
    A. They are caused by the lack of light reflected off the moon when the sun and the moon are on the opposite sides of Earth.
    B. They are caused by the changes in gravitational forces of the moon and the sun.
    C. They are caused by day and night.
    D. They are caused by the changes in seasons.

*The following question contains a blank marked "Select . . . ▼." Beneath it is a set of choices. Indicate the choice that is correct and belongs in the blank. (**Note**: On the real GED test, the choices will appear as a "drop-down" menu. When you click on a choice, it will appear in the blank.)*

52. If two planets of the same mass are orbiting a star at different distances, the planet that is closer to the star will travel [Select . . . ▼] the planet that is farther from the star.

Select . . . ▼
at the same speed as
faster than
more slowly than

Question 53 refers to the following passage.

The maximum size of a stable white dwarf star is about 1.4 times the mass of the sun. Stars with mass higher than this limit, called the Chandrasekhar limit, ultimately collapse under their own weight and become neutron stars or black holes. Stars with a mass below this limit are prevented from collapsing because pressure from their electrons prevents the atoms from collapsing.

53. What is the Chandrasekhar limit?

    A. the maximum mass that a black hole can contain
    B. the maximum density of a black dwarf star
    C. the maximum mass of a white dwarf star
    D. the minimum density of a black hole

54. What does NOT occur on the moon due to the lack of significant atmosphere?

    A. varying temperatures
    B. hills and valleys
    C. colorful sunrise and sunset
    D. rotation on its axis

Question 55 refers to the following passage.

Comets are small cosmic bodies that have an ellipsoidal orbit around the sun. They are composed of rock, dust, ice, and frozen gases. Comets do not become spherical because of their low mass and gravity. They are thought to be remnants of planetesimals, which are small bodies that were not large enough to become planets.

Many of the asteroids in the solar system are found in the asteroid belt. Asteroids are relatively small, inactive, rocky bodies orbiting the sun. Comets are active objects with ice that can vaporize in sunlight, forming an atmosphere of dust and gas and sometimes, a tail of dust and gas.

55. Which of the following statements about asteroids and comets is true?

    A. They orbit the sun.
    B. They eventually become planets.
    C. They both have ice as part of their composition.
    D. They both have metals as part of their composition.

*The following question contains a blank marked "Select . . . ▼." Beneath it is a set of choices. Indicate the choice that is correct and belongs in the blank. (**Note**: On the real GED test, the choices will appear as a "drop-down" menu. When you click on a choice, it will appear in the blank.)*

56. The three basic types of rock are sedimentary, metamorphic, and igneous. The differences among them have to do with how they are formed. Sedimentary rocks are formed from particles of sand, shells, pebbles, and plant and animal remains. Igneous rocks are formed from volcanic eruptions. As hot lava from within Earth cools, it hardens to rock. Metamorphic rocks are rocks that are buried deep within Earth and have changed composition over time due to extreme heat and pressure. Fossils are usually found in Select . . . ▼ rocks.

Select . . . ▼
igneous
metamorphic
sedimentary

57. The diagram below shows the sedimentary rock layers at Niagara Falls. Which type of rock layer appears to be most resistant to weathering and erosion?

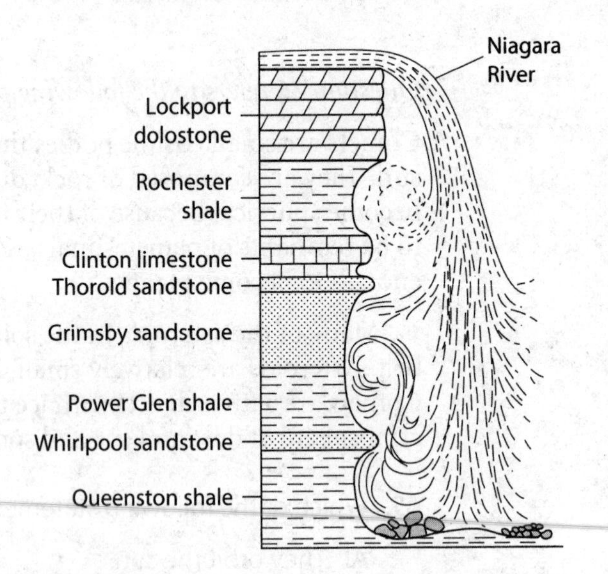

A. Power Glen shale
B. Lockport dolostone
C. Thorold sandstone
D. Rochester shale

58. According to the Big Bang theory, the relationship between time and the size of the universe from the beginning of the universe to the present is best represented by which of the following graphs?

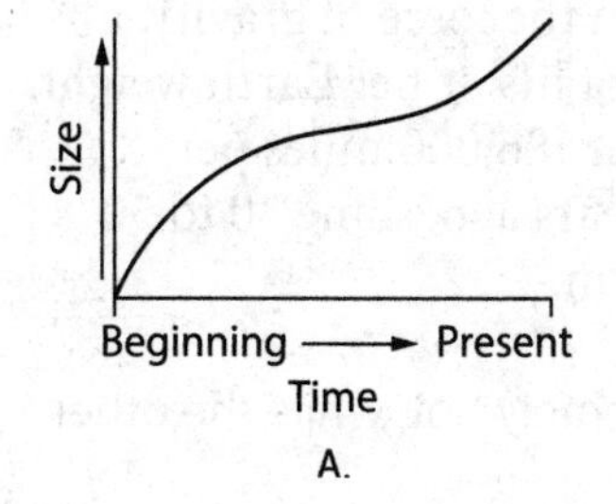

A.

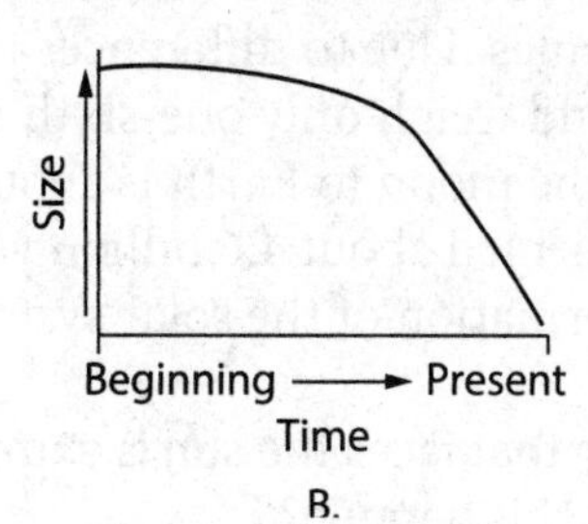

B.

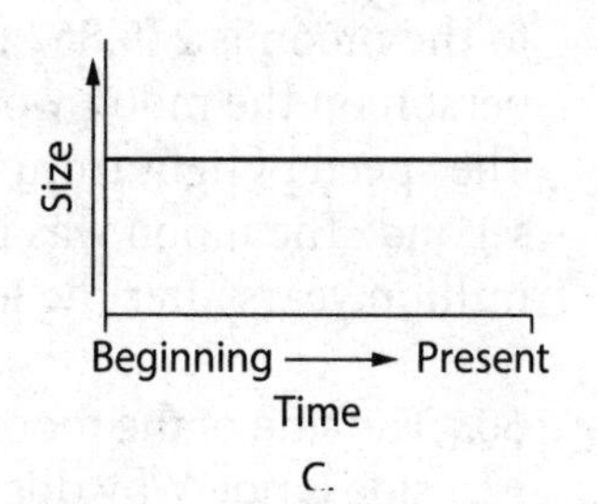

C.

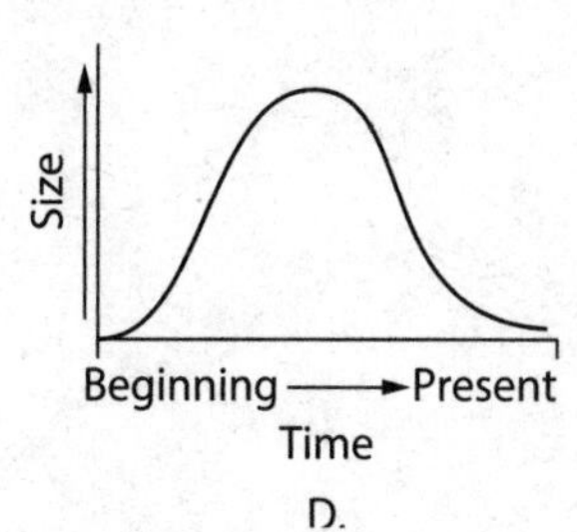

D.

59. The following diagram shows a large meteor impacting Earth's surface.

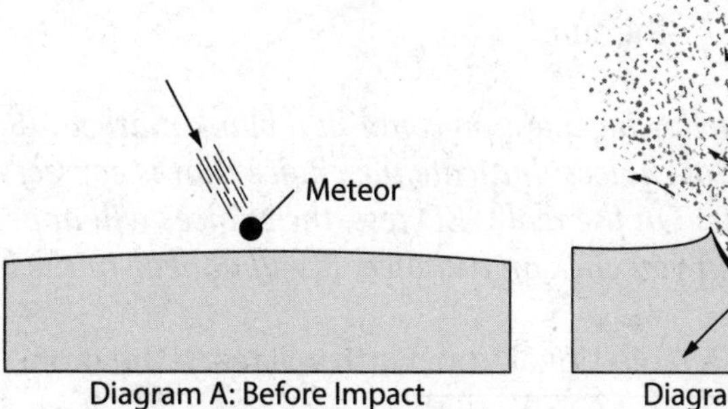

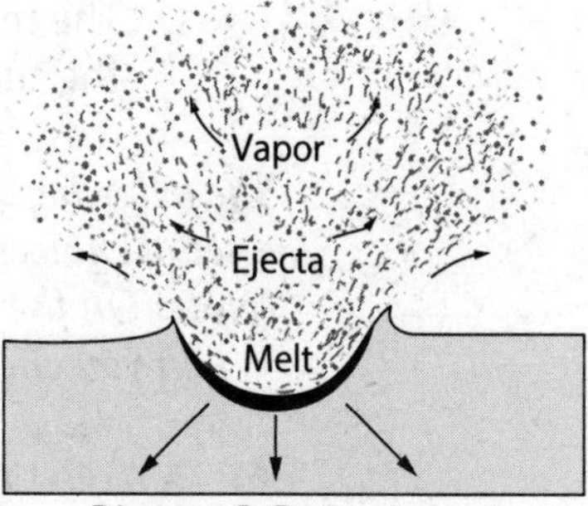

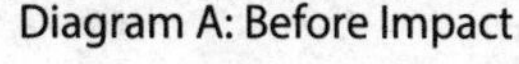
Diagram A: Before Impact

Diagram B: During Impact

The melt of the meteor is the material that melted during impact. The vapor is gases that formed. The ejecta are solid particles that were scattered into the air. Which statement best describes how the global climate could be affected by the meteor's impact?

A. The large quantities of ejecta could block sunlight, causing the global temperature to cool.
B. An increase in vapor and ejecta could reflect more sunlight from Earth, causing global temperature to cool.
C. Ejecta could settle into thick layers and absorb sunlight, causing Earth to heat up.
D. Forest fires produced from the impact could raise global temperatures.

Questions 60–62 refer to the following passage.

The moon is Earth's only natural satellite. It has a diameter of 2,160 miles, which is roughly 33 percent of Earth's diameter. The distance from Earth to the moon is 238,857 miles. Due to differences in the force of gravity, a person on the moon would weigh only one-sixth of his or her Earth weight. The speed of light from the moon to Earth is about 186,000 miles per second. The moon was formed about 4.6 billion years ago, some 30 to 50 million years after the formation of the solar system.

60. The side of the moon that faces the sun is extremely hot, while the other side is not. Why does this happen?

    A. The moon lacks an atmosphere to provide protection from the sun.
    B. The reflective rocks on the moon cause it to heat up.
    C. There is a lack of volcanic activity on the moon.
    D. The mineral composition of the moon is different on the side facing the sun.

*The following question contains a blank marked "Select . . . ▼." Beneath it is a set of choices. Indicate the choice that is correct and belongs in the blank. (**Note:** On the real GED test, the choices will appear as a "drop-down" menu. When you click on a choice, it will appear in the blank.)*

61. A radio signal from Earth will reach the moon in approximately Select . . . ▼ .

Select . . . ▼
1.15 seconds
2.15 seconds
3 seconds

62. If a person weighs 100 kg on Earth, his or her weight on the moon would be approximately

    A. 16 kg
    B. 20 kg
    C. 100 kg
    D. 600 kg

Write your answer in the blank.

63. The footprints left behind by the astronauts of *Apollo* will be visible for close to 10 million years because there is no ________________ on the moon to produce wind and rain.

64. Arches National Park in Utah is famous for its visually appealing natural arches. The salt beds deposited by a receding ocean created these arches. Once the salt shifted, the top layer started to be formed into domes and fins. Which of the following helped in creating these arches?

    A. extreme heat
    B. water and wind
    C. glaciers
    D. ocean tides

65. Describe the formation of a U-shaped valley.

Write your response on the lines.

____

____

____

____

____

____

66. Iridium is an extremely rare metal that is similar to platinum. Large concentrations of iridium are only found in meteorites (asteroids, meteors, or comets) that have crashed on earth or after large volcanic eruptions that thrust the iridium out from deep within the earth. Scientists concluded that a comet hit the earth and wiped out the dinosaurs because of the concentration of iridium found in rock stratum from 65 million years ago.

    Which of the following, if true, most strongly supports the scientist's conclusion?

    A. Dinosaurs were not the only animals to die out 65 million years ago.
    B. A comet struck the earth approximately 28.5 million years ago, but no widespread extinction resulted from that collision.
    C. A volcano large enough to generate high concentrations of iridium is extremely rare.
    D. Most scientists agree that a comet impact wiped out the dinosaurs.

Use the following information to answer Question 67.

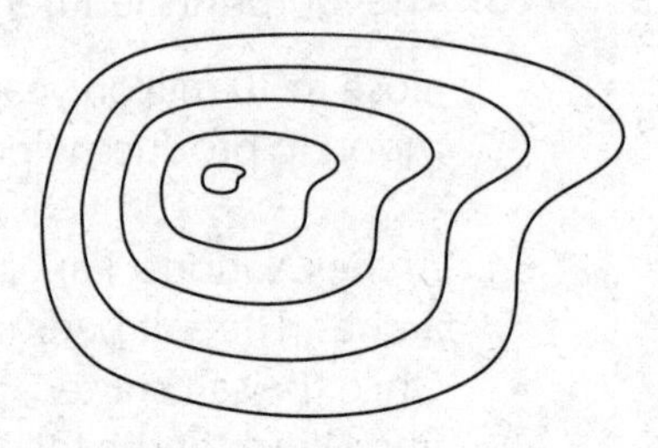

67. If the outer line on this topographical map is sea level and the contour interval is 20 meters, which of the following could be the height above sea level of the highest point shown on the map?

    A. 40 meters
    B. 70 meters
    C. 90 meters
    D. 120 meters

Use the following information to answer Questions 68–69.

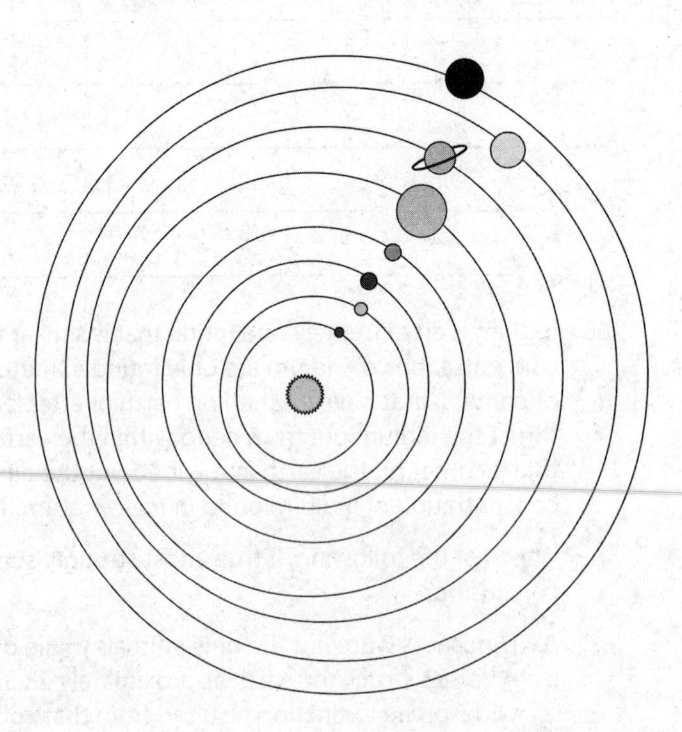

68. Which of the following planets were formed from the light gases of the outer solar nebula? *Place an X on the blank next to each of your choices.*

    __________ Mercury
    __________ Jupiter
    __________ Uranus
    __________ Mars
    __________ Neptune
    __________ Saturn
    __________ Venus

69. Which planet has the fastest orbit around the sun? *Write your answer in the blank.*

    ____

70. In which layer of the atmosphere does nearly all weather occur?

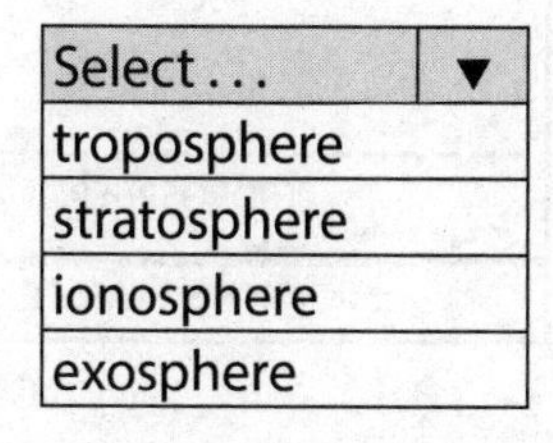

Use the following for Question 70.

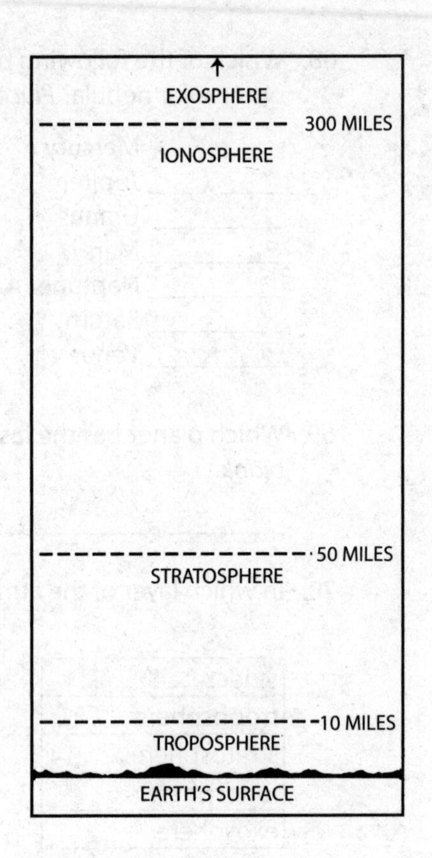

The following question contains a blank marked Select... ▼*. Beneath it is a set of choices. Indicate the choice that is correct and belongs in the blank. (**Note**: On the real GED test, the choices will appear as a "drop-down" menu. When you click on a choice, it will appear in the blank.)*

ANSWERS AND EXPLANATIONS

Chapter 1: Life Science

1. **A** From the charts, it can be seen that the amount of oxygen is less and the amount of carbon dioxide is more in exhaled air than in inhaled air.
2. Possible response: Fill a beaker with lime water. Blow exhaled air into the beaker with a straw. The lime water will become milky. This indicates that there is carbon dioxide in the exhaled air.
3. **D** This is an illustration of the human heart, the organ that pumps blood throughout the body.
4. **B** The enamel is the outer layer of the tooth.
5. **B** Teeth help in grinding food, thereby playing a major role in the digestive system.
6. **C** Different parts of the brain control different body functions.
7. **C** Sight is controlled by the occipital lobe and memory is maintained by the temporal lobe. To remember someone's face, coordination of these two lobes is important.
8. **B** The brain processes the sensory impulse from the hand and sends a message to react using motor nerves.
9. **sensory**
10. **B** The combined percent for bone and muscle is 59.7%. This is the greatest percent of the categories given in the graph.
11. **B** This is the only information that is supported by the graph.
12. **C** Women's bodies differ from men's bodies due to the presence of mammary glands and female reproductive organs, all of which contain fat-rich tissues.
13. **C** "Survival of the fittest" refers to the process of natural selection.
14. **A** A change to darker colored trees may lead to a decrease in the white moth population since the dark moths would be better protected from predators among dark-colored trees. This increases their chances for reproduction and passing their genes on to their offspring.
15. **C** Driving hybrid cars would not affect the amount of pollution coming from the factory.
16. **B** Valves keep blood flow moving in one direction.
17. **A** A bicycle tire also has valves to prevent the air from leaking out.
18. **closely** Species A and B are more closely related than are species A and C, as indicated by their proximity on the diagonal line.
19. **B** The oxygen consumption rate shown is between 0.2 and 0.4 liter per minute.
20. **D** This statement is the most complete summary of information provided by the graph.
21. **A** It can be inferred from the graph that the oxygen consumption reaches a steady-state rate, which is determined by the rate of exercise. Exercise at a more strenuous level increases the oxygen consumption rate.
22. **D** The oxygen consumption rate will remain steady at the now higher rate, so choice D is the only possible answer.
23. **D** Each jaw has 16 teeth.
24. **B** The shape of the teeth depends mostly on the type of food eaten.
25. **A** The smoker's heart would beat more rapidly to meet the body's oxygen needs. This is because the smoker's blood is not able to carry oxygen efficiently.
26. **B** This is an interesting fact, but it is not directly related to alcohol and health.

ANSWERS AND EXPLANATIONS

27. **A** This is the summary of the information contained in the first two paragraphs of the passage.

28. **B** Epilepsy cannot be passed on from one person to another.

29. **D** The experiment is designed to test how quickly a person reacts to the dropping ball.

30. **B** Alcohol interferes with both judgment and reaction time.

31. **D** Because candy is sweet in taste, Area 4 is where its taste is identified.

32. **positive** In this case, the platelets cling to the injured site to combat the bleeding and cause the attraction of even more platelets. Because the initial event is increased, it is positive feedback.

33. **negative** The insulin acts as a negative measure to counter the increased blood sugar level.

34. **C** Surrogate parenting is a product of progress made in the science of reproduction. Each other choice is a direct result of nature's design; they are processes that have not changed.

35. **C** According to the Gaia hypothesis, Earth can recover from things that happen because of features of Earth itself. Collision with an asteroid introduces something from outside Earth into Earth's system.

36. **C** Each other choice is mentioned in the third paragraph as a symptom of heat exhaustion. Below-normal temperature is also often a symptom.

37. **B** Lowering a victim's body temperature will help prevent the serious consequences of heatstroke, such as brain damage and death.

38. **D** Each of the other choices directly affects the exerciser's access to cool air and water.

39. **C** Air temperature is the least significant factor related to the victim's immediate health needs.

40. **B** This statement is the summary of the information provided in the chart.

41. **C** The need for vitamin supplements for anyone who eats a nutritious diet is controversial. Some medical experts recommend vitamin supplements; others do not.

42. Air-borne: sneezing, dusting. Blood-borne: infected syringe, open wound.

43. **D** Going to sleep early is a good habit, but this is not directly related to preventing the spread of diseases.

44. **A** Salted crackers are not a moist, high-protein food.

45. **B** Of the months listed, July is the month when most people will barbecue at home and have picnics and other outdoor gatherings.

46. **D** Of the choices given, the campfire is least likely to be a source of *Salmonella*.

47. **A** Hamburger buns are usually not moist, and they are not a high-protein source.

48. **D** Food is expected to be fresh and not contaminated at least until the sell-by date.

49. **B** This graph indicates that even as the death rate in the rest of the world is improving, it is not improving in sub-Saharan Africa.

50. **1900** This information is available from the graph.

51. **B** The diagram shows the levels of organization in a living organism.

52. **B** This is naturally acquired passive immunity because antibodies are passed from the mother to the infant through the mother's milk.

53. **D** Antibodies are produced in an animal and then injected into a person. This is an artificially acquired passive immunity.

54. **A** When a person is exposed to a disease, antibodies are produced. When the person

ANSWERS AND EXPLANATIONS

is exposed to the disease a second time, the number of antibodies is much greater. Because of this, the person usually does not get the disease a second time. This is a naturally acquired active immunity.

55. **C** Sunlight is necessary to conduct photosynthesis, so the plant in the dark room will not be able to prepare sugars.

56. **B** It is necessary to remove previously prepared sugars before starting the experiment.

57. **sugar; oxygen** These are the components to the right of the arrow in the equation.

58. **B** The sun is the primary source of energy in the ecosystem.

59. **C** This is an example of a food chain.

60. **C** Human lungs are moist and a constant source of oxygen.

61. **D** Keeping the planters on the same table would have little effect on plant growth. Each other choice plays an important role in the comparative growth of the plants.

62. **C** Planters A and C receive full days of sunlight but are given different amounts of water. These are the conditions being tested in this comparative study.

63. Independent: sunlight, water; Dependent: plant growth. Sunlight and water are the independent variables. Plant growth is dependent on both these factors, making it a dependent variable.

64. **B** Dental work often involves bleeding gums, so the choice is not a practical approach.

65. **C** The energy pyramid is not dependent on the number of organisms in each level, but rather on the energy flow between levels.

66. **D** From producers to primary consumers, 10% of 18,000, i.e., 1,800, units of energy get passed on. From primary to secondary consumers, 10% of 1,800, i.e., 180, units get passed on. Similarly, tertiary consumers receive 10% of 180, i.e., 18, units.

67. **B** Because there is huge loss of energy in each level, the number of organisms at higher levels is limited.

68. **secondary consumers** Carrion eaters consume the remains of dead animals.

69. **producer** Green plants start the food web because of their ability to make food.

70. **B** This statement is a brief and accurate summary of the passage.

71. **B** Cell walls give fruits, such as apples, their crunchy texture. A breakdown of cell walls results in a squishy texture.

72. **D** The misunderstanding of many people concerning food irradiation is not of direct interest to scientists trying to scientifically determine its safe use in food preservation.

73. **D** Sterilization is the only method that is considered practically 100 percent effective.

74. **A** Female condoms are not a very effective birth-control method, as the table shows, so an advertisement for female condoms would not likely feature this information.

75. **C** The best evidence would come from two groups of currently healthy people, one group that eats saccharin and one group that doesn't.

76. **parasitism** The tapeworms deprive the host of the nutrients.

77. **commensalism** The sharks are not affected by the fact that the remora fish eats up scraps of their food.

78. **mutualism** The bee gets nectar from the plant while helping pollinate the plant.

79. **B** Invertase is missing in the ants, so the acacia tree provides partially digested sugars.

ANSWERS AND EXPLANATIONS

80. **A** The extinction of similar mammals in Europe would have no effect on the well-being of animals in North America.
81. **C** This information is given in the first paragraph of the passage.
82. **D** This very serious condition is described in the third paragraph of the passage.
83. **C** All the other choices are possible results during pregnancy.
84. **B** A fetus receives all of its oxygen from the mother's blood. Smoking decreases the amount of oxygen available in this blood.
85. **A** Eukaryotic organisms have most of their DNA within the nucleus of cells.
86. **half as many** Daughter cells have half as many chromosomes as the parent cell after meiosis.
87. **cell wall** Because the cell wall is rigid in nature, it gives shape to the cell.
88. **A** The process of cell division gets initiated from the nucleus.
89. **C** Cells group together to form tissues. Tissues form organs. And several organs together form the organ systems.
90. **C** The order for mitosis is prophase, metaphase, anaphase, telophase.
91. **A** One cell divides into 2, 2 divide into 4, 4 into 8, and finally 8 divide into 16.
92. **B** The daughter cells in binary fission are an exact replica of the parent cell, so they cannot result in bio-diversity.
93. **C** $40 \times 12 \times 30 \times 100 = 1{,}440{,}000$ bacteria.
94. **B** The resulting cells of meiosis are haploid. So $46/2 = 23$.
95. **C** Mitosis produces 2 diploid cells. Meiosis produces 4 haploid cells.
96. **C** The reaction shows sunlight reacting with carbon dioxide and water to form glucose, a simple sugar that can be used to supply energy. This process is called photosynthesis.
97. **B** The rabbit and the eagles will not use photosynthesis to make their own food. Instead, they will use glucose from the rabbit in a process called cellular respiration to produce energy that can be stored in ATP molecules for later use.
98. **B** Aerobic organisms need to use oxygen gas in order to stay alive. Anaerobic organisms and anaerobic processes do not use oxygen.
99. **B** T will pair with A and C with G, so all other combinations are invalid.
100.

	s	s
S	**Ss**	**Ss**
s	**ss**	**ss**

101. **C** The two Ss offspring will have spherical seeds and the two ss offspring will have dented seeds.
102. **A** The dominance is due to the dominant trait and has no connection to the gender.
103. **D** All other options may cause mutation.
104. **C** Nothing is mentioned in the passage about the cost of medicines being tried with Alzheimer's patients.
105. **D** There is no sure cure for cancer, but much progress is being made toward the goal.
106. **C** The long-term effects of new cancer treatments, such as Imatinib, are not known; therefore, knowledge of long-term effects is one advantage of radiation treatment.
107. **C** The deer and the blackbird have differences in their food habits.

ANSWERS AND EXPLANATIONS

108. Possible response: The adult male orangutans of Borneo roam freely and do not stay close to any family unit. An adult male of Sumatra does stay close to his mate and offspring. On Sumatra, leopards prey on orangutan infants, hence the male is needed to protect the family. There is no such threat in Borneo. On Borneo, the male is not needed for protection and only increases competition for food.

109. **D** This information is given in the final paragraph of the passage.

110. **C** The passage states that the presence of the males increases competition for food.

111. **B** On each island, the orangutan has adapted to conditions in a way that best ensures the survival of the orangutan species.

112. **C** Producing the greatest number of surviving offspring is characteristic of all life forms.

113. **B** Charles Darwin was the famous naturalist whose theory of evolution is a landmark in scientific thought.

114. **A** Animals are generally free from predators when in captivity.

115. **finding** This discovery was a result of the experiment.

116. **experiment** This was the experiment itself.

117. **hypothesis** A possible explanation of the findings of the experiment.

118. **nonessential fact** Though an interesting fact, it is not directly related to the experiment.

119. **prediction** This may be discovered at some time in the future.

120. **A** By natural selection, mice with a color similar to that of the sand can survive and pass their genes on to future generations.

121. Possible response: The trout's population will be adversely affected by global warming and deforestation. Trout stay in cool parts of the water where there is more oxygen. Global warming would cause a rise in the temperature and deforestation would lead to fewer trees. The trout will no longer be able to reside in the shady parts of the pool.

122. **B** The trout's coloration blends with the natural surroundings, enabling it to fool the predators.

123. **A** In the chart, between years 3 and 6 the prey population (the gazelles) increases. Because of this increase in food, the predator population (the cheetahs) begins to thrive and grow in years 6 to 8.

124.

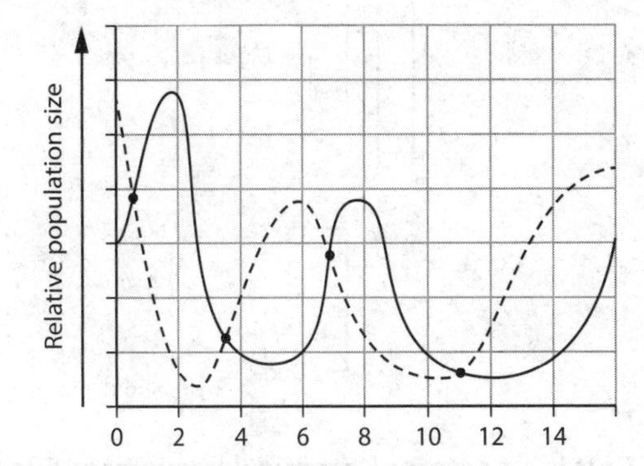

125. **D** Both the thermostat and the brain provide feedback so that the furnace and the body can make adjustments as needed. The adjustments made help maintain stability or homeostasis.

126. **D** In order for two organisms to synthesize similar enzymes, they must have DNA sequences that are similar, because DNA codes for the proteins that make up enzymes (and many other things). Because the dog and horse come from the same evolutionary line, they must have had a common ancestor that had a similar DNA sequence. That is why they will produce similar enzymes and other similar proteins as well.

127. **C** The gorilla and chimp are most directly related because they share a recent common ancestor.

ANSWERS AND EXPLANATIONS

128. Follow the lines from the cow and horse to see where they intersect.

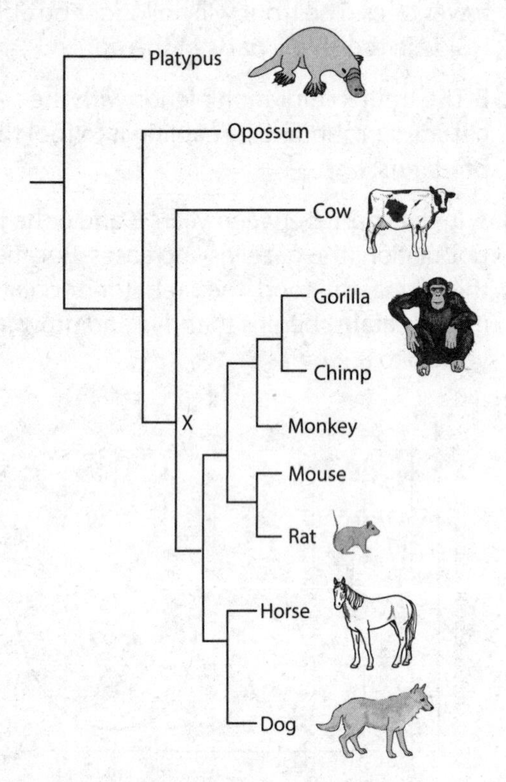

129. **Salt** is an external chemical barrier provided by the body's innate immune system. Mucus and skin are external barriers as well, but they are physical rather than chemical.

130. **C** Bones physically join with the neuromuscular system via connective tissues such as ligaments and tendons.

131. **D** The rough endoplasmic reticulum is shown. It can be distinguished from the smooth endoplasmic reticulum because it has little bumps on its surface. These are protein-manufacturing ribosomes.

132. **C** The cell membrane is selectively permeable, allowing it to control what enters or exits the cell.

133. **A** In Mendelian inheritance, the ratio of dominant to recessive traits in a dihybrid cross is 9:3:3:1.

134. **C** Choice C shows a decline in one population disrupting its equilibrium with other populations.

135. **A** The dark moths have survived to pass on their genes, which increases their population while the light moth population decreases. This is an example of natural selection.

136. **Gel electrophoresis** is a laboratory method used to separate and analyze mixtures of DNA, RNA, or proteins according to molecular size and charge.

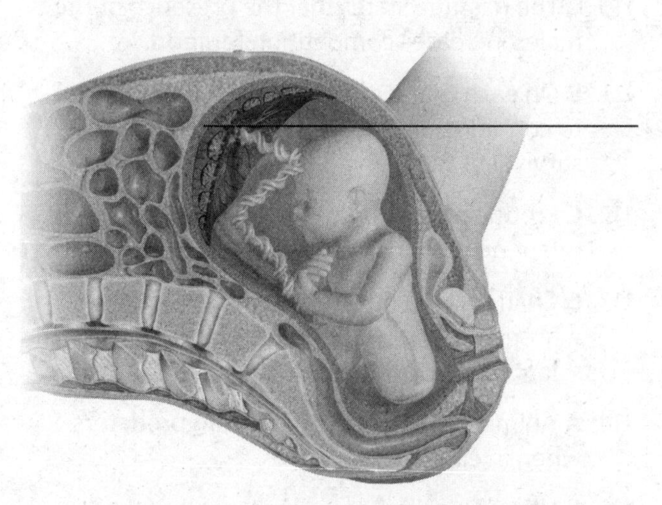

137. The placenta is an organ that connects the developing fetus to the uterine wall to allow nutrient uptake, thermo-regulation, waste elimination, and gas exchange via the mother's blood supply.

138. **A** The umbilical cord carries oxygenated blood to the fetus.

139. **B** The most energy produced is from the lower level of the pyramid, by the autotrophs.

140. **D** The link between photosynthetic organisms and carnivores is herbivores and a rabbit is an herbivore. Chickens are omnivores.

ANSWERS AND EXPLANATIONS

Chapter 2: Physical Science

1. **C** The flow of heat is from higher temperature to lower temperature.

2. **B** Beaker X contains 70 mL of water at a temperature of 80°C. A higher temperature means a higher average kinetic energy.

3. **not flow from the iron ball to the water or from the water to the iron ball** Since both the iron ball and water are at the same temperature, there is no flow of heat.

4. **B** Use the relationship between Fahrenheit and Celsius temperature: $y°F = 9/5\,x°C + 32$, if $x°C = y°F = T$, then

   $T = 9/5\,T + 32$

   $9/5\,T - T = -32$

   $T = -40$

5. **C** Water boils at 212°F and at 100°C.

6. Box A: −273°C; −459°F; 0 K

   Box B: +273 K; 32°F; 0°C

7. **C** As $y°F = \frac{9}{5}x°C + 32$, if $y°F = -22$, then

   $-22 = 9/5\,x°C + 32$

   $9/5\,x°C = -22 - 32$

   $x°C = -30$

8. **B** Kelvin scale has no negative numbers, as absolute zero 0 K is the lowest possible temperature for any substance.

9. **C** As $C = \Delta Q / (m \times \Delta T)$

   $128\ \text{J/(kg·K)} = \Delta Q / [225\ \text{g} \times (25 - 15)]$

   $\Delta Q = [0.128\ \text{J/(g·°C)} \times 225\ \text{g} \times 10°\text{C}]$

   $\Delta Q = 288\ \text{J}$

10. **D** Per Table 1, aluminum has a higher specific heat of 902 J/(kg·K) than copper 385 J/(kg·K). Thus the metal with the lower specific heat will have greater temperature change. This is because, according to the following formula, the change in temperature is inversely proportional to the specific heat.

    Since $C = \frac{\Delta Q}{(m \times \Delta T)}$, if Q and m ($m = 1$ kg in this case) are kept constant, $C \propto \frac{1}{T}$.

11. Possible response: Lake Michigan is a body of water with a large mass and a large specific heat (*C*). It would take a lot of solar energy (heat) absorption to increase its temperature from the cold wintry temperatures to the higher summertime temperatures. It may take a couple of months of summer to heat the large mass of water.

12. **A** The temperature recorded for liquid A after 6 minutes is 50°C; therefore,

    $\Delta T = T_{final} - T_{initial} = 90°C - 50°C = 40°C$

13. **liquid B** The lower the specific heat, the greater is the change in the temperature. Because liquid B undergoes a greater change in temperature, it has the lower specific heat.

14. **B** Given $m = 50$ g, $c = 4.18$ J/(g·°C), $T_{water} = 88.6°C$, $T_{final} = 87.1°C$,

    $\Delta T = T_{final} - T_{water} = -1.5°C$

    Solve for Q_{water}:

    $Q_{water} = m_{water} \times C_{water} \times (T_{final} - T_{water})$

    $(50.0\ \text{g}) \times [4.18\ \text{J/(g·°C)}] \times (-1.5°\text{C})$

    $= -313.5\ \text{J}$

    (The negative sign indicates that heat is released by the water.)

    Now, at thermal equilibrium,

    Q_{water} released $= Q_{metal}$ absorbed

    Therefore, $Q_{metal} = +313.5$ J (The positive sign indicates that heat is absorbed by the metal.

    $Q_{metal} = m_{metal} \times c_{metal} \times (T_{final} - T_{metal})$

ANSWERS AND EXPLANATIONS

$313.5 \text{ J} = 12.9 \text{ g} \times c_{metal} \times (87.1°\text{C} - 26.5°\text{C})$

$c_{metal} = 0.40103 \text{ J/(g·°C)}$ (rounded)

$= 401.03 \text{ J(kg·K)}$

15. **temperatures** When two bodies are in thermal equilibrium, they have the same temperature.

16. **B** 20 J. $\Delta U = Q - W$

    $= W = Q - \Delta U = 60 - 40 \text{ J} = 20 \text{ J}$

17. **B** Given that the work done by the athlete = 650 kJ and heat released = 425 kJ, as $\Delta U = Q - W$, then,

    $\Delta U = -425 - 650 \text{ kJ} = -1{,}075 \text{ kJ}$

    (heat released, so negative Q)

18. **A** Point 1 is B.

    Point 2 is C.

    Point 3 is A.

    From C → A, net work is zero, which means there is no change in volume with change of pressure.

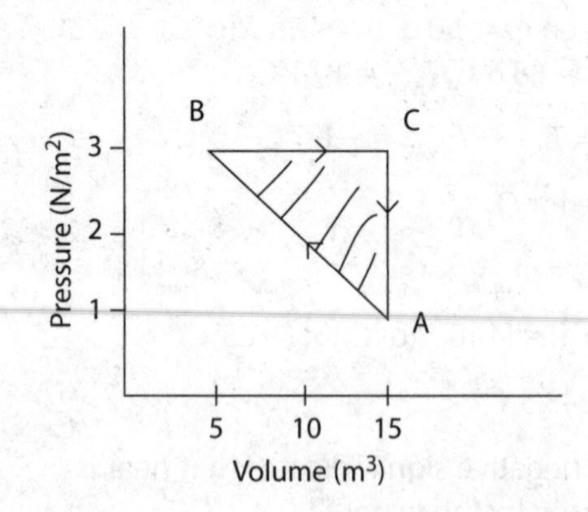

19. **D** $W = P(V_2 - V_1)$

    Here, $V_2 = 2V_1$

    So $W = P \times (V_2 - V1) = P \times (2V_1 - V_1)$

    $= P \times V_1 = 151{,}500 \text{ Pa} \times 4 \text{ m}^3$

    $= 606{,}000 \text{ J}$

20. **D** $\%e = \frac{W}{Q_H} = \frac{15\text{J}}{60\text{J}} = 25\%$

21. **B** 2,500 J is the work output. Then the work input (X) can be calculated as

    30% of $X = 2{,}500$ J

    $X = 2{,}500/0.30 = 8{,}333$ J (rounded)

    Now the released heat energy is the difference between input and output heat energy, that is, $(8{,}333 - 2{,}500) \text{ J} = 5{,}833 \text{ J}$.

22. **D** Heat energy transferred from the movement of fluid is called convection.

23. **C** All matter is made up of particles called molecules and atoms. These particles are always in different types of motion—translation, rotational, and vibrational. These motions give the particles kinetic energy. Temperature is a measure of the average amount of kinetic energy possessed by the particles in a sample of matter. The more the particles vibrate, translate, and rotate, the greater the temperature of the object. Heat is a flow of energy from a higher temperature object to a lower temperature object.

24. Possible response: Heat energy first got transferred from the stove burner in direct contact with the pot by conduction. Then heat got transferred through water by convection and finally through air again by convection.

25. **A** For the first few seconds, the temperature outside of the air-conditioned bus is greater than the temperature of the glass that was still at the same temperature as the inside of the air-conditioned bus. Thus the water vapor in the surrounding air condenses to form small water droplets that make the glass fog. As the temperature of the glass increases to match the outside temperature, the water droplets evaporate and the glass becomes clear again.

26. **melting; sublimation** When a solid changes into a liquid, the process is called melting, and when a solid changes directly into a gas, the process is called sublimation.

27. **C** A gas has no definite volume.

ANSWERS AND EXPLANATIONS

28. **endothermic** Ice cream melting on a hot, sunny day is an example of an endothermic reaction (a system that absorbs heat from the surroundings).

29. **hot; release** An exothermic reaction releases heat, and thus the test tube where the reaction is taking place becomes hot.

30. **exothermic; released** An exothermic reaction releases heat.

31. Possible response: To sustain life, we require energy. We obtain this energy from our food. In the process of digestion, food molecules get converted into simpler substances, such as glucose. Then glucose combines with oxygen and provides energy to our body. The whole process is known as respiration. Energy is released during the process of respiration, so it is considered an exothermic reaction. For example,

    $C_6H_{12}O_6 + 6O_2 \rightarrow 6CO_2 + 6H_2O + \text{energy}$

32. **equal to** While doing work, an object loses its potential energy, but the same is gained in the form of kinetic energy and vice versa, and thus the total mechanical energy is always conserved.

33. **B** Heating the reactants, using a catalyst and higher concentrations of reactants would all increase the amount of ammonia.

34. **A** In order for the hydrogen and nitrogen to react, there must be effective collisions between the hydrogen and nitrogen gases.

35. **D** 68 grams

36. **B** $K_E = ½ \times mv^2 = ½ \times 625 \text{ kg} \times (18.3 \text{ m/s})^2 = 1.05 \times 10^5 \text{ J}$

37. **D** A flowing river is a water body in constant motion; hence, it possesses kinetic energy.

38. **quadrupled** The kinetic energy equation, $K_E = ½mv^2$, reveals that the kinetic energy of an object is directly proportional to the square of its speed, so if K_E is doubled, its speed will increase four times.

39. Possible response: A hammer is a tool that utilizes mechanical energy to do work. The mechanical energy of a hammer gives its ability to apply a force to a nail to cause the nail to be displaced. Because the hammer has mechanical energy (in the form of kinetic energy), it is able to do work on the nail.

40. **A** Light can act like both a wave and a particle.

41. **A** In hydroelectric dams, water from an elevation (possessing gravitational potential energy) is made to fall from a height (kinetic energy), generating electrical energy.

42. **C** The battery converts chemical energy to electrical energy, and the electrical energy converts to light energy.

43. **natural gas** Natural gas when burned has the least emissions and causes less pollution, so it is more environmentally friendly.

44. Possible response: Oil, natural gas, and coal are called *fossil fuels* because they have been formed from the organic remains of prehistoric plants and animals.

45. **B** Natural gas produces less CO_2 than coal. CO_2 is the major contributor of greenhouse gases.

46. **nuclear energy** This is obtained by nuclear fission, which is the process of splitting apart uranium atoms in a controlled manner that creates energy.

47. Geothermal sources: volcano, hot springs, natural geysers. Biomass sources: trees, crops, plants.

48. **fusion** Nuclear fusion is the process that causes the sun to radiate energy in the form of light and heat commonly known as solar energy.

49. **biomass** The chart shows that 49% of the renewable energy used in the United States is biomass. This is a larger percent than for the other renewable sources.

50. **D** The answer is deduced from the graph, which shows the total fossil fuel consumption

ANSWERS AND EXPLANATIONS

to be the sum of coal (18%) + petroleum (36%) + natural gas (27%) = 81% of total fossil fuel consumption.

51. **D** We absorb the least amount of sunlight by wearing light-colored clothes and staying out of direct sunlight.

52. **D** The burning of fossil fuels releases carbon dioxide into the atmosphere, which traps heat close to Earth's surface.

53. **D** When released, CFC causes more heat to be trapped in the atmosphere, contributing to global warming.

54. **A** Frequency refers to the number of occurrences of a periodic event over time and is measured in cycles per second. In this case, there is one cycle per 2 seconds, so the frequency is 1 cycle/2 s = 0.5 Hz.

55. **B** The wavelength must be 8 m (see diagram). The period is 3 seconds, so the frequency $(f) = 1/T = 1/3 = 0.333$ Hz.

    Now use $v = f \times \lambda$.

    $v = (0.333 \times 8) = 2.67$ m/s

56. **D** The particles would be moving back and forth in a direction perpendicular to energy transport. The waves are moving westward, so the particles move northward and southward.

57. **A; D** The wavelength is the distance from crest to crest (or from trough to trough or between any two corresponding points on adjacent waves). The amplitude is the distance from rest to crest or from rest to trough of a wave.

58. **C** Choices A, B, and D are all examples showing how light can act as a wave. Because of the wave-particle dual nature of light, light can also act as a particle. The sensor is one example of a device that uses the photoelectric effect of light acting like particles called photons.

59. Longitudinal waves: sound waves, compression, rarefaction. Transverse waves: light waves, crest, trough.

60. **B** Each element on the periodic table has a specific atomic number that matches the number of protons in the nucleus of an atom of the element. The number of electrons in an atom of the element often matches the number of protons, but the number of electrons can also be different because an atom can gain or lose electrons to form an ion.

61. **D** Substance D, with the highest melting point, will need the most average kinetic energy (temperature) to break its bonds. Therefore, substance D on the table has the strongest bonds.

62. **C** Substances made of all nonmetals will have lower melting points. For example, wax is made up of elements that are nonmetals. That is why a wax candle melts easily with a small flame. In contrast, iron, which is all metal, needs a blow torch to melt it.

63. Speed: 0.5 m/s (The physics teacher walked a distance of 12 m in 24 s; thus her average speed was 0.50 m/s.)

    Velocity: 0 m/s (Because her displacement is 0 m, her average velocity is 0 m/s. Displacement refers to the change in position.)

64. **B** Velocity = total displacement/time = 12,250 m/3,600 s = 3.40 m/s

65. **zero** Because resting state is zero velocity, the initial velocity of a body starting from rest is zero.

66. **C** $p = m \times \Delta v = m \times (v_2 - v_1)$

    $p = 1 \text{ kg} \times [(-2 \text{ m/s}) - 4 \text{ m/s}]$

    $\Delta p = -6$ kg m/s

    v_2 is negative here as the ball bounces back striking the wall.

67. **less** momentum of bullet = 0.004 kg × 400 m/s = 1.6 kg·m/s

    and

    momentum of baseball = 0.150 kg × (90 × 1,000 m/3,600 s) = 0.150 kg × 25 m/s = 3.75 kg·m/s

ANSWERS AND EXPLANATIONS

68. **C** Momentum is directly proportional to the product of an object's mass and velocity.

69. Possible response: The hose is pushing a lot of water (large mass) forward at a high speed. This means the water has a large forward momentum. In turn, the hose must have an equally large backward momentum, making it difficult for the firefighters to manage.

70. **A** By the law of conservation of momentum, $[(m_1 \times v_{1a}) + (m_2 \times v_{2a})]_{before} = [(m_1 \times v_{1b}) + (m_2 \times v_{2b})]_{after}$.

    $v_{1b} = [1 - (0.5 - 1.33)]$ m/s $= 0.335$ m/s

71. **B** The velocity–time graph reveals that the object starts with a zero velocity (as read from the graph) and finishes with a large, negative velocity; that is, the object is moving in the negative direction and speeding up. An object that is moving in the negative direction and speeding up is said to have a negative acceleration.

72. **C** Acceleration of an object depends on its speed and direction.

73. **B** Acceleration = velocity/time = (80/0.2) m/s^2 = 400 m/s^2

74. **B** Inertia is the book's resistance to a change in its state of motion. The book remained at rest and did not move forward with the car.

75. Possible response: When the ball leaves the rim, the ball follows path 2. Once leaving the rim, the ball will follow path 2, which is an "inertial path" (i.e., a straight line). At the instant shown in the diagram, the ball is moving to the right; once leaving the rim, there are no more unbalanced forces to change its state of motion. Paths 1 and 3 show the ball continually changing its direction once leaving the rim.

76. **B** Two forces of equal magnitude and opposite direction will cancel each other out.

77. **A** The distance–time graph where distance is parallel to the time axis shows that the body is at rest.

78. **C** Newton's first law of motion states that an object continues to be in its state of rest or motion unless it is acted on by an external force. When the moving horse suddenly stops, the lower part of the body of the rider contacts with the horse's back and suddenly comes to rest, while the upper part of the body tends to retain its state of motion due to inertia. As a result, the rider is thrown forward.

79. **A** $a = F/m$ and $a = v/t$

    $F/m = v/t$, $v = (400\text{ N}/0.5\text{ kg}) \times 5\text{ s} =$ 4,000 m/s

80. **D** The mass of the cars does not change, so $F1 = m \times 1.8$ m/s^2 and $F2 = m \times 1.2$ m/s^2.

    $F1/F2 = 3{:}2$

81. **three** Since acceleration is directly proportional to the net force applied to an object, if the net force is tripled, the acceleration will be tripled.

82. **equal to** According to the given situation, the velocity of the train is constant; so acceleration is also constant and the masses of the cars are the same. Therefore, the forces acting on the first and the last car of the train must be equal.

83. When the diver jumps off a diving board, the board springs back and forces the diver into the air. The action force exerted on the board by the diver causes a reaction force by the board on the diver. The force of the diver on the board is equal and opposite to the force exerted by the diving board on the diver.

84. **different** The action and reaction forces act on different bodies.

85. **D** Electrons form the least part of the mass of an atom.

86. **B** $P = W/t = (0.50\text{ J}) / (2.0\text{ s}) = 0.25\text{ W}$

87. Possible response: Randy and Aaron do the same amount of work. They apply the same force to lift the same barbell the same distance above their heads. Aaron delivers more power

ANSWERS AND EXPLANATIONS

since he does the same work in less time. Power and time are inversely proportional.

88. **A** $P = W/t = (m \times g \times h) / t$

$2{,}000\ W = [(m \times 10 \times 10) / 60]$

$m = 1{,}200$ kg

89. **C** The two energies are equal at the halfway point between 0 and 10.

90. **D** Only option D is not true.

91. **knife** A knife is an example of a wedge. It changes the direction of the force by spreading apart the object being cut.

92. **C** The mother is using an inclined plane.

93. **wheel and axle** The handle of the screwdriver is the wheel and the shaft is the axle.

94. **D** When you use an inclined plane to raise an object, you do the same total amount of work as when you lift the object directly. However, you use less force, even though you move the object a longer distance.

95. **A** Mechanical advantage = effort arm / load arm = 100 cm / 20 cm = 5

96. **4.2** Mechanical advantage = length of ramp divided by height. 5.3 / 1.26 = 4.2

97. **A** Protons have a positive charge.

98. **protons; electrons** In a neutral atom, the numbers of protons and electrons are equal.

99. **proton** The number of protons distinguishes one element from another.

100. **C** Calculate the charge on the ion:

Neutral magnesium has 12 protons: 12 + and 12 electrons: 12–, so the charge was 0 initially; after magnesium loses two electrons, the total number of electrons becomes 10–.

Therefore, the net charge = (12+) – (10–) = 2+; Mg^{2+}.

101. **20** An element's atomic number is the number of protons in its nucleus.

102. **C** A neutral atom must have an equal number of protons and electrons.

103. **C** In an isotope of argon, the atomic mass number = total number of protons + neutrons in a nucleus = 40, and an atomic number = total number of protons in the isotope = 18. Therefore, the total number of neutrons = 40 – 18 = 22.

104. **B** ^{35}Cl and ^{37}Cl have the same atomic number but different mass numbers.

105. **B** A reaction that releases heat is an exothermic reaction.

106. **C** Hydrogen has the least atomic weight, and lighter gases travel faster.

107. Physical properties: density, color, viscosity. Chemical properties: flammability, reactivity; acidity.

108. **physical; chemical** The density of iron is a physical property and the process of rusting is a chemical change in the iron.

109. **D** Change in color is a change in a physical property.

110. **C** The chemical nature of water is not changed. There is a phase change from liquid to gas, but steam can be converted to water again by condensation.

111. **B** Dew is formed when water vapor in the air condenses on a cool surface to form drops of water.

112. **D** When energy is extracted from the atmosphere to evaporate liquid water, the atmosphere will cool down.

113. **C** Stronger molecular bonds are the reason why honey is so thick at room temperature.

114. **B** Density = mass/volume = 42 g/15 mL = 2.8 g/mL

115. **B** The density of a liquid determines whether it will float on or sink in another liquid. A liquid will float if it is less dense than the liquid it is placed in.

ANSWERS AND EXPLANATIONS

116. **C** Density is a scalar quantity as it is dependent on mass and volume; vector quantities need to have a direction.

117. **D** There are six atoms of oxygen on both sides of the reaction; $3 \times 2 = 6$ atoms as reactants and $2 \times 3 = 6$ atoms as products.

118. **D** There is an equal number of atoms on both sides of the reaction.

119. **plus sign; arrow** *Reacts with* is symbolized by a plus sign and *yields* is symbolized by an arrow.

120. **coefficient** This is the number in front of an element or compound in a chemical equation.

121. **C** $CaCO_3$ undergoes decomposition by heat.

122. **bases; salt** In a neutralization reaction, an acid reacts with a base to form water and salt.

123. **B** In a combination reaction, reactants combine to form one product, and there is no unreacted product present in the reaction.

124. **more** Oxygen and other gases are more soluble at colder temperatures than at warmer temperatures.

125. **solvent** A solute is a substance in lesser quantity, dissolved in another substance, known as a solvent, in greater quantity to make a solution.

126. **D** Nitrogen is a gas, so increasing pressure and lowering temperature increase solubility.

127. **D** Use the formula $F = \frac{9}{5}C + 32$. Because the Celsius temperature is 25 degrees, plug that into the formula for C. $F = \frac{9}{5} \times 25 + 32$. Now solve for *F*.

$$F = \frac{9}{5} \times 25 + 32$$
$$F = 45 + 32$$
$$F = 77$$

128. **B** Two atoms are said to be isotopes when they have different numbers of neutrons but the same number of protons.

129. **B** In the process of diffusion, a substance moves from an area of high concentration to areas of low concentration.

130. **B** Neutral pH measures 7 on the pH scale.

131. **B** An ionic bond is usually between a metal and a nonmetal. The two substances bond by attracting oppositely charged ions.

132. **C** Ca and Sr are in the same family: alkaline earth metals. They have the same number of valence electrons.

133. **B** Argon and the other noble gases have the most chemically inert atoms because they have full valence shells with no electrons that need to seek bonding.

134. **B** Synthesis of water is shown by combining two hydrogen and one oxygen atoms: $2H_2 + O_2 \rightarrow 2H_2O$

135. **D** The independent variable is what is changed in the experiment. In this case, that is the amount of time each student spends studying. The dependent variable is the scores the students make on the test. The subject matter of the test and the number of questions on the test are controls because they would be the same for everyone.

136. **C** A shift toward the blue end of the spectrum indicates shorter wavelength. The star is moving closer to the observer.

137. **D** Nitrogen is a gas, so increasing pressure and lowering temperature increase solubility.

138. **B** Of the four choices, only choice B is reversible (the products re-form the reactants) and can reach equilibrium. Each of the other choices is a reaction that only makes the products and is not reversible.

139. **B** There are drawbacks to using nuclear reactions to produce energy that is then used to heat steam and produce electricity. However, no greenhouse gases are produced when nuclear reactions are used to generate electricity, and that is a benefit.

140. **D** The triple point is a certain temperature and pressure at which a substance can exist in all three phases of matter.

ANSWERS AND EXPLANATIONS

Chapter 3: Earth and Space Science

1. **water** The water cycle is the journey water takes as it circulates within Earth and the atmosphere and back again.
2. **C** Transpiration is the process by which water evaporates from inside leaves to the atmosphere.
3. **D** After water droplets have condensed to make clouds, they join together to form larger droplets. When these drops are heavy enough to fall, they return to Earth's surface as rain, hail, or sleet. If clouds are made of ice particles instead of water drops, they can produce snow. This process is called precipitation.
4. **D** In the carbon cycle, fossil fuels are produced from the decay of organisms that were once living.
5. **C** Carbon dioxide from the atmosphere is used by plants, which along with sunlight produce carbohydrates for energy through photosynthesis.
6. **carbon dioxide** Plants use atmospheric CO_2 and sunlight to build carbohydrates by photosynthesis. Deforestation leads to a decrease in plant population that results in increasing CO_2 content in the atmosphere.
7. **decaying plants** Bacteria use the nitrogen compounds from decaying plant and animal matter.
8. 1, 4, 3, 5, 2 is the correct order for the steps in the nitrogen cycle.
9. **C** Wood is not a form of hydrocarbon.
10. **C** As the trees and plants died, they sank to the bottom of the swamps, oceans, or any water body. They formed layers of a spongy material called peat. Over many hundreds of years, the peat remained covered by sand and clay and other minerals, which turned into a type of rock called sedimentary. More and more rock piled on top of more rock, and it weighed more and more as it pressed down on the peat. The peat was squeezed and squeezed over millions of years and turned into coal, oil or petroleum, and natural gas.
11. **they take a very long time to form** Fossil fuels are considered nonrenewable because they are being extracted and consumed much faster than they can be formed. It takes millions of years to turn peat into fossil fuels.
12. **A** Alternative energies that are renewable, such as solar, wind, and hydroelectric energy, are all methods for generating energy without burning fossil fuels and creating more carbon dioxide.
13. **B** The end product of carbon dioxide mixing with rain water is acid rain. Acid rain can lower the pH of a stream or lake, making it more acidic. This can have a harmful impact on wildlife. Acid rain can also damage structures built with marble.
14. **D** The acids form high in the atmosphere and can be carried by wind and weather fronts.
15. **C** Valleys are shaped by weathering. A V-shaped valley is shaped by the movement of running water. A U-shaped valley is shaped by the movement of a glacier.
16. **D** Weathering can break down rocks. The smaller, weathered rocks are called fine sediments.
17. **A** The hypocenter or focus is the underground point where an earthquake rupture starts. The epicenter is the point on Earth's surface vertically above the hypocenter, the point in the crust where a seismic rupture begins.
18. **A** From the passage, the epicenter is directly above the focus and the fault scarp is the feature that looks like a step.

ANSWERS AND EXPLANATIONS

19. **more** A higher number on the modified Mercalli scale indicates more structural destruction.

20. **Richter; modified Mercalli**

21. **C** Hurricanes gather heat and energy through contact with warm ocean waters. Evaporation from the seawater increases their power. The center of these winds is a low pressure area, so wind with higher pressure all around rushes toward the center.

22. A. eye; B. spiral rain bands; C. eye wall; D. counterclockwise rotation.

23. **Atlantic; Pacific**

24. **C** Hurricanes can cause a huge water surge and massive flooding. A dike is an embankment of soil, rock, and other materials to prevent such water surges from causing floods.

25. **B** While Earth can be at a range of distances from the sun during the course of a year, it is not this distance that determines the seasons. Instead, it is the amount of direct sunlight that an area on Earth receives that determines the seasons. That amount of sunlight is determined by the tilt of the planet.

26. **A** While estimates vary, it is believed that eventually Earth's human population will reach the maximum number that the planet's resources can sustain. This number is called Earth's carrying capacity.

27. **C** The "hot thin soup" concept describes early Earth's atmosphere, which is thought to have contained gases made out of carbon, oxygen, hydrogen, and nitrogen. Gases containing these elements were able to react to form larger organic molecules. Those are called the "molecules of life." They are DNA, fats, proteins, and sugars. Glucose is a simple form of sugar.

28. **D** A red shift in the light spectrum of a star indicates that the star is moving away from Earth.

29. Possible response: There would be extreme fire all around. Even the smallest fires would take the form of extremely devastating explosions because oxygen is highly combustible.

30. **D** Temperature increases in the stratosphere because there is a layer of concentrated ozone, which absorbs UV rays and warms the air.

31. **C** Planes fly in the stratosphere because there are no weather hindrances and the air is thinner.

32. **5°C** The average decrease shown in the chart is 5°C.

33. **7,000 feet** According to the information in the table, the temperature will be approximately 0°C at 7,000 feet.

34. **D** There are many causes of skin cancer; UV rays are just one of them.

35. **B** Increasing the number of trees would decrease global warming by absorbing carbon dioxide and producing oxygen.

36. **D, A, F, E, B, C** Letter A shows evaporation; Letter B shows condensation; Letter C shows transpiration; Letter D shows precipitation; Letter E shows surface runoff; and Letter F shows infiltration.

37. **D** The sun is fueled by a series of fusion reactions.

38. **C** Because the planets have cyclic, elliptical orbits, their positions can easily be predicted by humans (sometimes with a little help from computer models). While there are warning systems in place to predict volcanic eruptions and earthquakes, those events do not take place in regular cycles. Finally, asteroids and meteorites and other space objects orbit around the sun, but many are too small to detect until the very last moment before they collide with Earth, making them very unpredictable.

ANSWERS AND EXPLANATIONS

39. Physical: ocean waves, rain, frost. Chemical: acid rain, carbonic acid, oxygen. Organic: earthworm, lichens, moss.

40. **A** Wind-blown sand moving close to the desert surface forms mushroom rocks.

41. **D** Movement of the continents is best related to plate tectonics caused by convection within Earth.

42. **B** When magma rises and forms a ridge in the middle of an ocean, it pushes the neighboring continents away from each other.

43. **C** The amount of aluminum (7%) is one-fourth that of silicon (28%).

44. **destructive plate** Ocean trenches such as the Mariana Trench are formed by destructive plate boundaries.

45. **lava; magma**

46. **B** The Mid-Atlantic Ridge lies along the boundary line of the two tectonic plates. The outpouring of magma and the presence of volcanoes along this ridge are evidence of this boundary.

47. **C** Plate tectonics is the movement of the continental plates due to convection within Earth's mantle layer. The arrows in the diagram show that the Rift region in eastern Africa (the crosshatched region) is moving eastward.

48. **B** At one time, the continents were located next to each other and have slowly spread apart over time. The ridge has most likely been midway between the continents during all this time.

49. **red** Light from objects that are moving away from Earth produces spectral lines that have a red shift.

50. **Big Bang** The theory of the origin of the universe is known as the Big Bang theory.

51. **B** The gravitational forces of the sun and the moon cause and affect tides.

52. **faster than** The planet that is closer to the star will travel faster.

53. **C** Stars with a mass greater than the Chandrasekhar limit will collapse into black holes or neutron stars.

54. **C** Sunrise and sunset as we know them on Earth are due to Earth's atmosphere.

55. **A** The only statement about comets and asteroids that is true is that they both orbit the sun.

56. **sedimentary** Sedimentary rocks are formed by deposits over time, and fossils are often found in their layers.

57. **B** All of the rock layers in the diagram have been impacted and rounded out by the falling water except for the topmost layer made of Lockport dolostone. Because that layer retained its shape, it is the most resistant to weathering and erosion.

58. **A** The universe has been expanding since the Big Bang. The only graph that shows continual expansion of the universe is choice A.

59. **A** The ejecta from a meteor impact can create tremendous dust clouds that can travel several miles into the air. These dust clouds can block sunlight from reaching Earth's surface, causing global temperatures to fall.

60. **A** When sunlight hits the moon's surface, the temperature can reach 253°F. The side of the moon not facing the sun has temperatures dipping to −243°F.

61. **1.15 seconds** It is the same as the speed of light.

62. **A** One-sixth of 100 kg is ≈16 kg.

63. **atmosphere** Due to the lack of an atmosphere, there is no erosion on the moon and imprints stay for a very long time.

64. **B** Over long periods, rain, ice, and wind create arches.

ANSWERS AND EXPLANATIONS

65. Possible response: Some features on Earth are shaped by water and ice. This process is called weathering. Erosion caused by slow-moving glaciers cut through rocks to form U-shaped valleys.

66. **C** If high levels of iridium are only caused by volcanic eruption or by meteorites, and volcanic eruptions of that magnitude are rare, then the likelihood of the iridium being deposited by a meteorite increases. Choice C best strengthens the scientists' argument.

67. **C** If the contour interval is 20 meters, then the innermost ring is 80 meters. The area inside it cannot be more than 100 meters, so 90 meters is a possible measurement.

68. **Jupiter**, **Saturn**, **Uranus**, and **Neptune** The four gas giants (Jupiter, Saturn, Uranus, and Neptune) formed from gases in the outer nebula, with cores made of ice, rock, and metal flakes.

69. **Mercury** The planet closest to the sun has the fastest orbit around it. Mercury orbits the sun at 47.87 km/s.

70. **Troposphere** Almost all of Earth's weather takes place in the lowest level shown, the troposphere.

POSTTEST

Science

40 questions | **90 minutes**

This posttest will give you an opportunity to evaluate your readiness to take the GED Science test. It includes topics from the three science disciplines represented by the three chapters in this workbook: Life Science, Physical Science, and Earth and Space Science.

Try to answer every question, in a quiet area so you are free from distractions and with enough time. The usual time allotted for the test is 80 minutes. Remember that it is more important to think about every question than it is to finish ahead of time. Answers can be found at the end of the posttest.

Directions: Answer the following questions. For multiple-choice questions, choose the best answer. For other questions, follow the directions preceding the question.

1. Energy transfer as heat between two objects depends on which of the following?

   A. the difference in mass of the two objects
   B. the difference in volume of the two objects
   C. the difference in temperature of the two objects
   D. the difference in composition of the two objects

Write your answers in the blanks.

2. The energy transferred as heat is always directed from an object at ______________ temperature to an object at ______________ temperature.

POSTTEST

Questions 3 and 4 refer to the following.

You are aware that Earth rotates and makes one complete rotation every day. But did you know that the sun also rotates? Scientists observed this by marking the position of sunspots, which appear dark. By tracking the movement of the sunspots, scientists can measure the rotation of the sun. Different parts of the sun actually rotate at different paces. At its equator, the sun rotates once every 25 days (Earth days), and at its poles, it rotates every 36 days.

3. In a complete Earth year, approximately how many times will the sun rotate at its equator?

   A. 12 times
   B. 15 times
   C. 10 times
   D. 22 times

*The following question contains a blank marked "Select . . . ▼." Beneath it is a set of choices. Indicate the choice that is correct and belongs in the blank. (**Note:** On the real GED test, the choices will appear as a "drop-down" menu. When you click on a choice, it will appear in the blank.)*

4. The sun is orbited by Select . . . ▼.

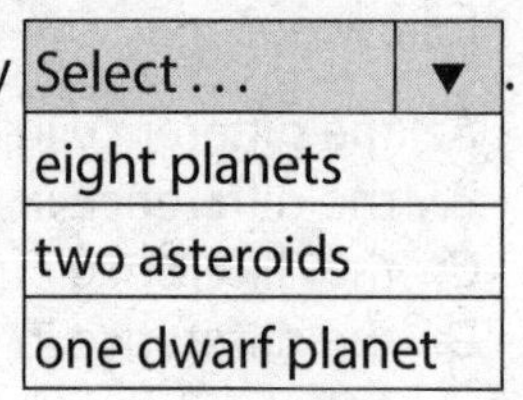

POSTTEST

Question 5 refers to the following passage.

The human body is an excellent example of coordination between organ systems. Each and every organ system works in perfect harmony with other systems to ensure the smooth functioning of the entire body. For example, human cells require oxygen to carry out metabolic activities. Cells break down food in the presence of oxygen to release energy and carbon dioxide. The air breathed in reaches the lungs; from there, the pulmonary veins absorb a part of the oxygen and carry the oxygen-rich blood to the heart. The heart then pumps the blood throughout the body, where the oxygen is absorbed and carbon dioxide is released back into the blood vessels. This blood is carried back to the lungs, from which the carbon dioxide is released back to the air when the person breathes out.

Write your answers in the blanks.

5. The example in the passage shows coordination between the ________________ and ________________ systems.

6. Mercury-in-glass thermometers work on the principle of volumetric expansion, which is that most materials expand as they get warmer. A common glass thermometer contains liquid mercury that rises upward in a glass tube as the temperature of the thermometer increases. The liquid moves up the tube for which of the following reasons?

   A. Only the glass contracts as its temperature decreases, so the liquid mercury moves up the tube to occupy more of the tube's internal space.
   B. While both the glass and the liquid mercury contract as their temperatures increase, the glass contracts more rapidly than the mercury does.
   C. The pressure of gas above the liquid mercury decreases as its temperature rises, and the liquid mercury is sucked upward in the glass tube.
   D. While both the glass and the liquid mercury expand as their temperatures increase, the liquid mercury expands more rapidly than the glass does.

POSTTEST

7. The DNA of three species of plants is sequenced using a gel electrophoresis. The results of the process show the following:

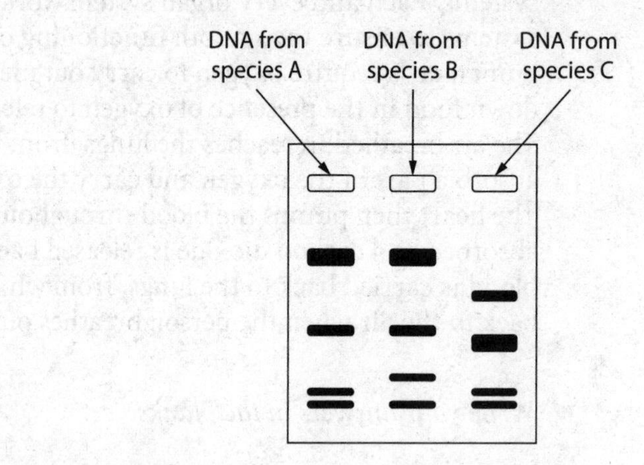

Which two species of plant are most closely related?

A. A and B
B. A and C
C. B and C
D. There is not enough conclusive evidence to determine which two are most closely related.

8. The following diagram shows two types of bacteria being exposed to an antibiotic.

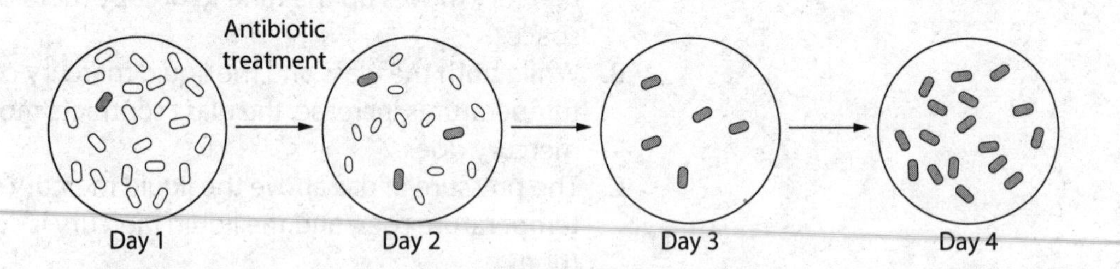

By day 4, one of the strains of bacteria was able to thrive and continue to reproduce. Which of the following reasons would explain why this took place?

A. None of the bacteria were able to survive the exposure to the antibiotic.
B. The bacteria found a new food source, making them invincible to the antibiotic.
C. Some of the bacteria already had the genes that made them resistant to the antibiotic.
D. Bacteria can instantly change their DNA sequence to become more resistant to antibiotics.

POSTTEST

Refer to the following information to answer Question 9.

Homeostasis refers to the ability of cells or body systems to keep a stable internal environment as the external environment changes. Cells and body systems monitor their internal environment and adjust the internal conditions if a change occurs. These adjustments may be made through a negative feedback system.

A **negative feedback** system is a process in which an initial change will bring about an additional change in the opposite direction.

A **positive feedback** system is a process in which an initial change will bring about an additional change in the same direction. Positive feedback does not maintain homeostasis.

9. Which of the following is NOT an example of a negative feedback system?

   A. A person is hungry for a long time, so that person's body metabolism slows down, allowing the body to survive on less food.
   B. The kidneys detect a decrease of oxygen in the blood, so they secrete a hormone called erythropoietin that increases the production of red blood cells.
   C. Blood pressure increases, so the brain signals the heart to slow down its rate, thus reducing the pressure.
   D. Proteins are only partially digested, so the digestive system secretes hydrochloric acid and pepsin, speeding up the process.

POSTTEST

Use the following information to answer Question 10.

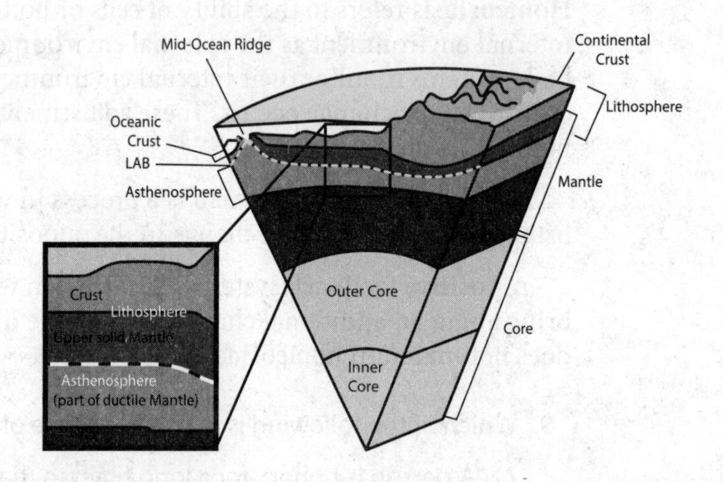

*The following question contains a blank marked "Select . . . ▼." Beneath it is a set of choices. Indicate the choice that is correct and belongs in the blank. (**Note:** On the real GED test, the choices will appear as a "drop-down" menu. When you click on a choice, it will appear in the blank.)*

10. The liquid part of Earth's surface can be found in the 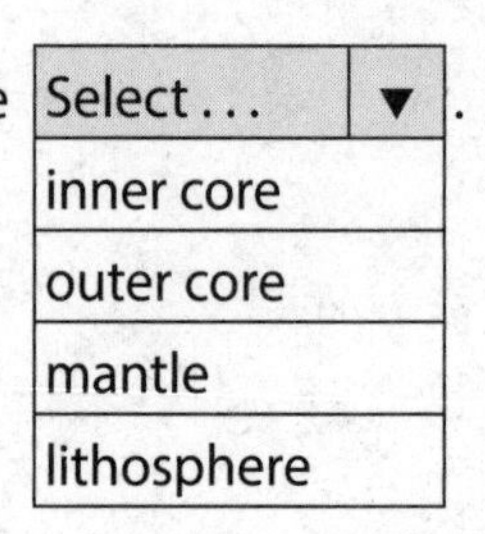.

POSTTEST

Questions 11 and 12 refer to the following chart.

Process	Constant	By First Law of Thermodynamics
Isobaric	Pressure	$\Delta U = Q + W$
Isochoric	Volume	$\Delta U = Q$
Isothermal	Temperature	$\Delta U = 0$
Adiabatic	No heat exchanged	$\Delta U = W$

ΔU = change in internal energy; Q = heat exchanged; W = work done

*The following question contains two blanks, each marked "Select . . . ▼." Beneath each one is a set of choices. Indicate the choice that is correct and belongs in the blank. (**Note:** On the real GED test, the choices will appear as a "drop-down" menu. When you click on a choice, it will appear in the blank.)*

11. Based on the chart, choose the correct term for each of the following pressure (P) vs. volume (V) graphs.

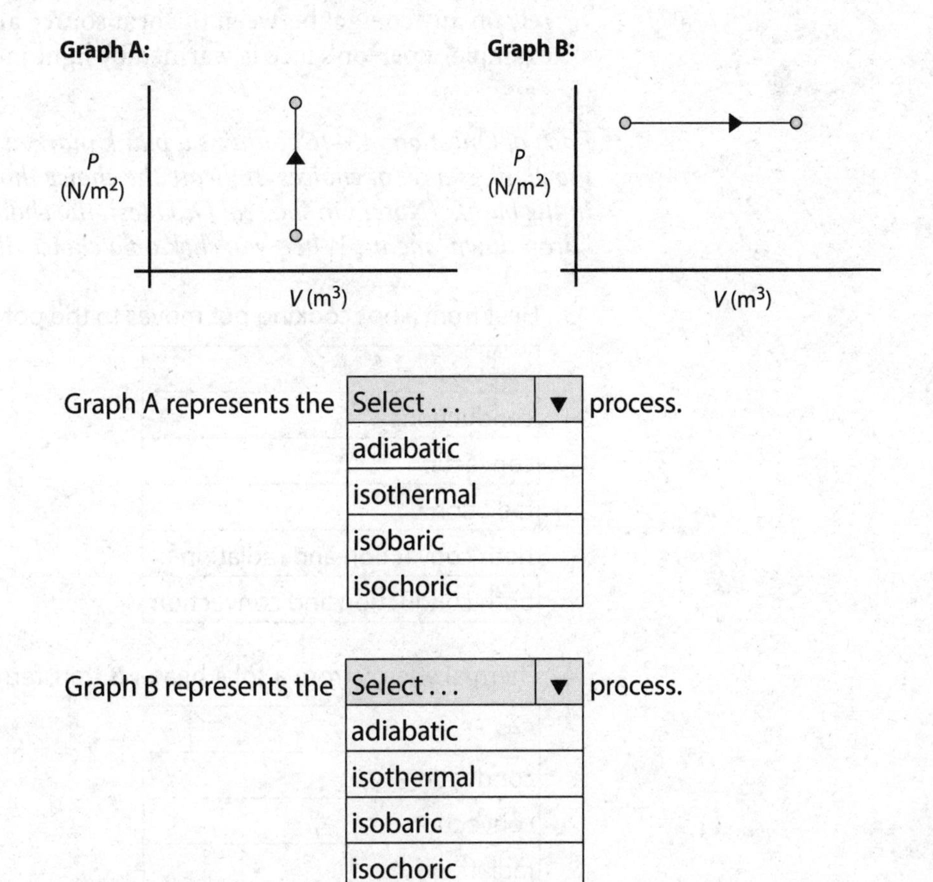

POSTTEST

Write your answers in the blanks.

12. During an isothermal process, both ______________ and volume change, but ______________ remains constant; therefore, the change in internal energy is ______________.

Questions 13–16 refer to the following information.

- **Conduction**—The transfer of heat or thermal energy by diffusion and collisions of particles within a substance due to a temperature gradient. By this process, heat is transferred through one substance to another when the two substances are in direct contact. For example, heat is transferred to a metal pot placed on a hot burner.
- **Convection**—The transfer of heat from one place to another by the movement of fluids. For example, when air heats up, the particles move farther apart and become less dense, causing the heated air to rise.
- **Thermal radiation**—The transfer of heat by radiation generated by the motion of charged particles. This method of heat transfer does not rely on any contact between the heat source and the heated object. For example, a person's face is warmed by light radiated from the sun.

*Each of Questions 13–16 contains a blank marked "Select . . . ▼." Beneath the blank is a set of choices. Indicate the choice that is correct and belongs in the blank. (**Note:** On the real GED test, the choices will appear as a "drop-down" menu. When you click on a choice, it will appear in the blank.)*

13. Heat from a hot cooking pot moves to the pot handle by

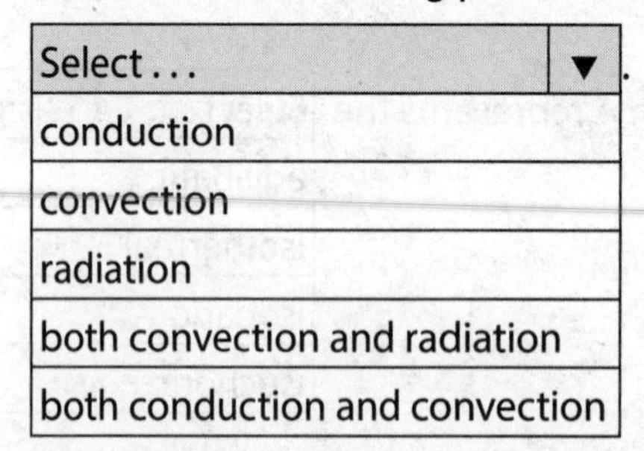

Select . . . ▼
conduction
convection
radiation
both convection and radiation
both conduction and convection

14. Thermal energy from a solar heater is transferred to a person's body by

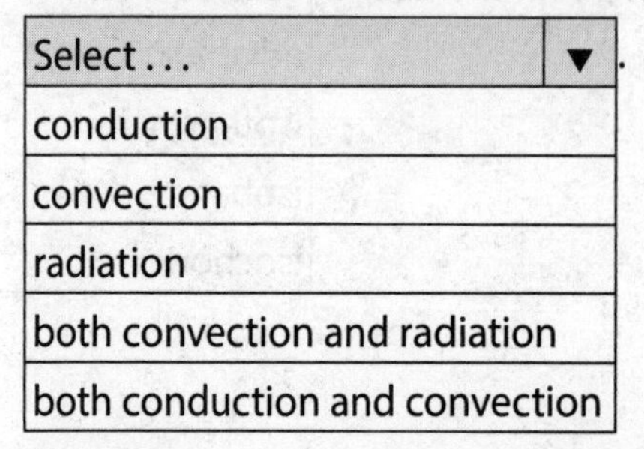

Select . . . ▼
conduction
convection
radiation
both convection and radiation
both conduction and convection

POSTTEST

15. Thermal energy from a fire moves up through a chimney by

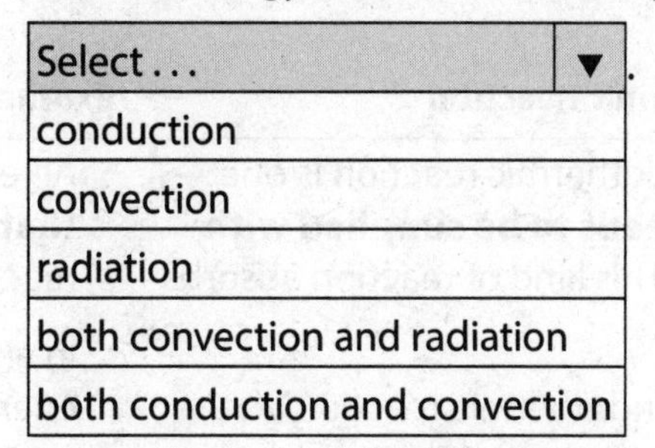

16. Thermal energy in a hot iron is transferred to a shirt by

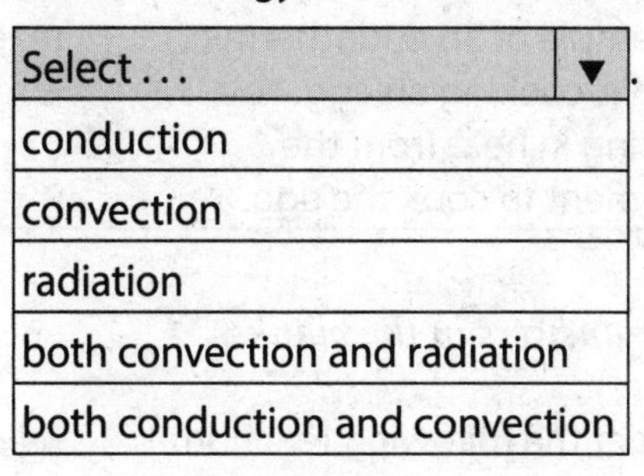

POSTTEST

Question 17 refers to the following chart.

Endothermic Reaction	Exothermic Reaction
• An endothermic reaction is one that **needs to be supplied with heat.** This kind of reaction absorbs heat. • In an endothermic reaction, the energy content of the reactants is less than that of the product. • One example of an endothermic reaction is cooking an egg. You must bring in heat from the environment to cook the egg.	• An exothermic reaction is one **that produces heat.** This kind of reaction releases heat. • In an exothermic reaction, the energy content of the products is more than that of the reactants. • One example of an exothermic reaction is a fire in a fireplace; heat is released from the burning logs.

Write your answers in the blanks.

17. Consider the following reaction:

    $4Al(s) + 3O_2(g) \rightarrow 2Al_2O_3(s) + \text{energy}$

    This reaction is an ________________ reaction. Heat is ________________ by the reaction.

18. According to the law of conservation of energy, energy can neither be created nor be destroyed, only transferred and transformed from one form to another. What kind of energy conversion is taking place when an electric bread toaster is switched on?

    A. Chemical → Electrical
    B. Electrical → Thermal
    C. Electrical → Sound
    D. Electrical → Chemical

19. Nuclear energy is obtained by nuclear fission, which is the process of splitting apart uranium atoms in a controlled manner that creates energy. This is quite dangerous because if the chain reaction of splitting the atoms is not controlled very carefully, an atomic explosion could occur. The fission process gives off heat energy, which is used to boil water in a power plant's reactor core. The steam created with this water is used to turn a turbine, generating electricity. Nuclear energy is derived from

    A. combustion of atoms of U-235
    B. fission of atoms of U-235
    C. fusion of atoms of U-235
    D. the breaking of U-235 bonds

POSTTEST

Question 20 refers to the following data.

Carrying capacity is the maximum number of individuals that can be sustained in a given environment at a given point of time.

The following graph shows the effect of an overshoot of population on the carrying capacity of an area.

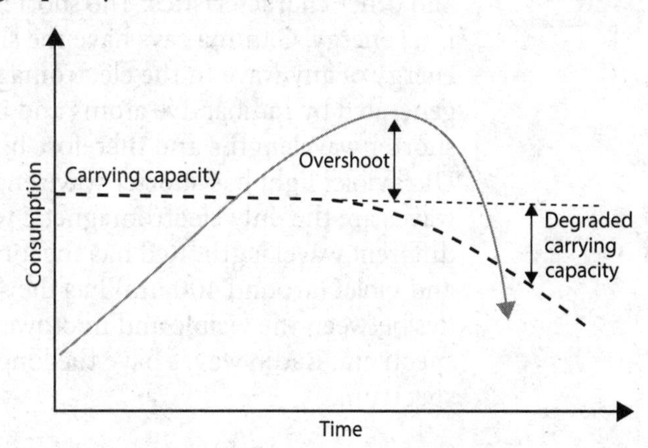

20. Which of the following situations best describes the graph?

    A. The overshoot of population never reached the carrying capacity.
    B. The overshoot of population had no effect on the carrying capacity.
    C. The population never stopped increasing, even after carrying capacity was reached.
    D. The overshoot of population caused the carrying capacity to decrease.

Question 21 refers to the following data.

The wave equation states the mathematical relationship between the speed (v) of a wave and its wavelength (λ) and frequency (f).

$$\text{speed} = \text{wavelength} \times \text{frequency}$$

$$v = \lambda \times f$$

21. A sound source sends waves of 400 Hz. It produces waves of wavelength 2.5 m. The velocity of the sound waves is

    A. 100 m/s
    B. 1,000 m/s
    C. 10,000 m/s
    D. 3,000 km/s

POSTTEST

Question 22 refers to the following passage.

Electromagnetic waves are produced by the vibration of charged particles and do not require a medium through which to travel. The electromagnetic spectrum is divided into several regions based on frequencies, wavelengths, and other characteristics. The shorter the wavelength of a wave, the greater is its energy. Gamma rays have the shortest wavelengths and the most energy of any wave in the electromagnetic spectrum. These waves are generated by radioactive atoms and in nuclear explosions. X-rays have shorter wavelengths and therefore higher energy than ultraviolet waves. Ultraviolet light has shorter wavelengths than visible light. Visible light waves are the only electromagnetic waves we can see. Each color has a different wavelength. Red has the longest wavelength (around 700 nm), and violet (around 400 nm) has the shortest wavelength. Infrared light lies between the visible and microwave portions of the electromagnetic spectrum. Radio waves have the longest wavelengths in the electromagnetic spectrum.

22. The diagram represents regions of the main electromagnetic spectrum in terms of decreasing wavelength and increasing energy. Which answer choice correctly labels regions 1, 2, 3, and 4?

1	Infra-red	*2*	*3*	*4*	Gamma rays

A. 1 = radio waves; 2 = ultraviolet rays; 3 = visible light; 4 = X-rays
B. 1 = radio waves; 2 = visible light; 3 = ultraviolet rays; 4 = X-rays
C. 1 = visible light; 2 = ultraviolet rays; 3 = X-rays; 4 = radio waves
D. 1 = visible light; 2 = ultraviolet rays; 3 = radio waves; 4 = X-rays

POSTTEST

Question 23 refers to the following information.

The following is a chart of information recorded over a period of 10 years. It shows the relationship between a predator (tiger) population and a prey (deer) population.

Year	No. of Deer	No. of Tigers
2001	100	8
2002	123	12
2003	200	18
2004	175	15
2005	130	11
2006	90	6
2007	110	8
2008	150	14
2009	180	16
2010	160	13

*The following question contains two blanks, each marked "Select... ▼." Beneath each one is a set of choices. Indicate the choice that is correct and belongs in the blank. (**Note:** On the real GED test, the choices will appear as a "drop-down" menu. When you click on a choice, it will appear in the blank.)*

23. The number of predators Select... ▼ as the number of prey

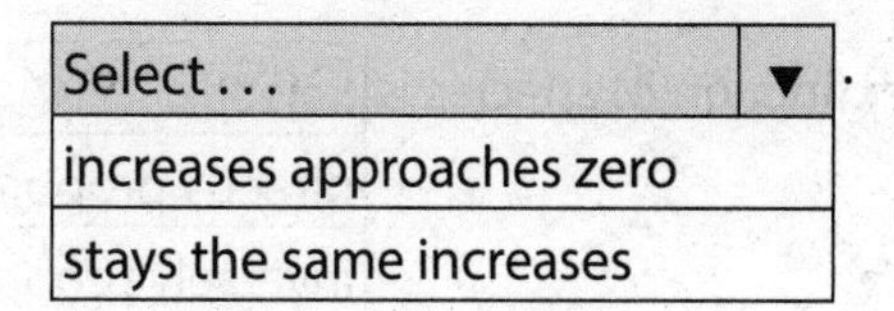

.

POSTTEST

Question 24 refers to the following information.

The following graph shows the extinction of species and the increase in human population.

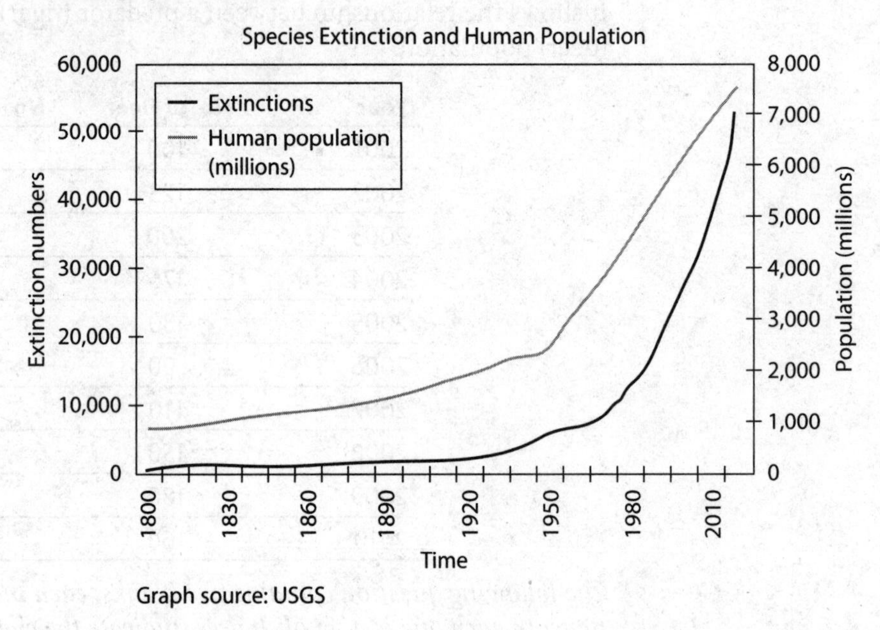

Graph source: USGS

*The following question contains a blank marked "Select . . . ▼." Beneath it is a set of choices. Indicate the choice that is correct and belongs in the blank. (**Note:** On the real GED test, the choices will appear as a "drop-down" menu. When you click on a choice, it will appear in the blank.)*

24. The extinction of species is Select . . . ▼ the increase in human population.

- directly proportional to
- inversely proportional to
- unaffected by

POSTTEST

Question 25 refers to the following information.

A group of cells working together is called a tissue. The human body contains four main types of tissues:

- **Epithelial tissue**—Cells that cover and protect underlying tissue
- **Nervous tissue**—Cells that carry nerve signals throughout the body
- **Muscle tissue**—Cells that contract and relax for the movement of skeletal parts or in the functioning of organs
- **Connective tissue**—Cells that join, support, cushion, and nourish organs

25. Indicate the correct box where each of the following belongs.

- Biceps
- Skin
- Cartilage
- Brain cells
- Ligament
- Cardiac muscle

A: Epithelial tissue	B: Nervous tissue	C: Muscle tissue	D: Connective tissue

POSTTEST

Question 26 refers to the following passage.

Newton's second law states that the acceleration of an object is dependent on the object's mass and the amount of force applied on the object. Therefore,

acceleration (*a*) = force (*F*) / mass (*m*)

Acceleration (a) is the rate of change of velocity (v) of an object. Therefore,

acceleration = velocity/time

$$a = v/t$$

The unit of acceleration is m/s^2. The unit of force is newton (N) and 1 newton = kg (m/s^2). Any object usually has more than one force acting on it at any time (e.g., driving force, air resistance, and friction). If two forces are acting in the opposite direction, then the difference between those two forces gives the resultant force.

26. What is the acceleration (a) of a car with a mass of 1,000 kg if the driving force is 3,600 N and the friction/air resistance is 800 N?

A. 2.8 m/s^2
B. 28 m/s^2
C. 3.6 m/s^2
D. 0.8 m/s^2

*The following question contains a blank marked "Select . . . ▼." Beneath it is a set of choices. Indicate the choice that is correct and belongs in the blank. (**Note:** On the real GED test, the choices will appear as a "drop-down" menu. When you click on a choice, it will appear in the blank.)*

27. Water and other substances enter the cell through diffusion. Diffusion is the process of flow of a substance from an area of higher concentration to an area of lower concentration. The cells in the root of the plant are responsible for the diffusion of water from the soil into the cell, which is then transported up the plant. To show this, an experiment was conducted in which a white flower was placed in water mixed with red coloring. After a few hours, it was observed that red streaks started appearing in the flower. For water to flow into the flower's cells, the concentration of water in the cells must be Select . . . ▼ the concentration of water in the container.

Select . . . ▼
less than
more than
equal to

POSTTEST

28. Renewable energy resources are sources of power that quickly replenish themselves and can be used again and again. Similarly, nonrenewable resources are those sources of power that cannot be used over and over again. Indicate the correct box where each of the following energy sources belongs.

- Solar
- Wind
- Coal
- Geothermal
- Oil
- Wood

A: Renewable sources of energy	B: Nonrenewable sources of energy

POSTTEST

Question 29 refers to the following passage and figure.

While working on a cladistics project, students learned that the first amniotic egg evolved approximately 340 million years ago in ancestral reptiles. The evolution of the amniotic egg allowed reptiles to expand into drier terrestrial habitats. The students represented this and other information they learned with a cladogram, as shown below.

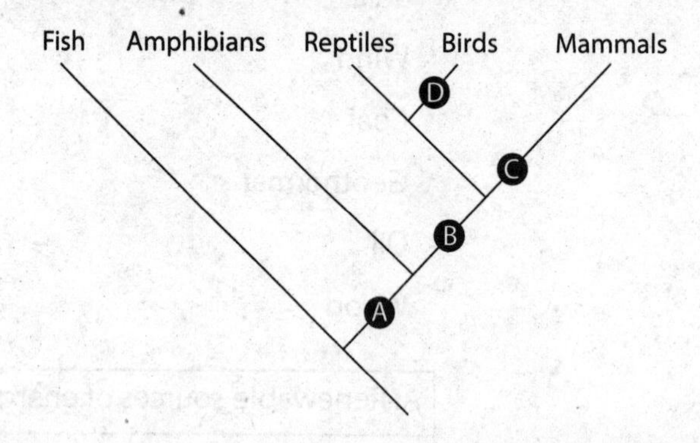

*The following question contains a blank marked "Select . . . ▼." Beneath it is a set of choices. Indicate the choice that is correct and belongs in the blank. (**Note**: On the real GED test, the choices will appear as a "drop-down" menu. When you click on a choice, it will appear in the blank.)*

29. Based on the information above, Select . . . ▼ on the cladogram

Select . . . ▼
point A
point B
point C
point D

corresponds with the appearance of the amniotic egg.

POSTTEST

Questions 30 and 31 refer to the following passage.

Meiosis is the type of cell division in which the parent cell divides to form four daughter cells, each of which has half as many chromosomes as the parent cell.

Mitosis is the type of cell division in which the parent cell divides to produce two identical daughter cells.

*Questions 38 and 39 each contain blanks marked "Select ... ▼." Beneath each blank is a set of choices. Indicate the choice that is correct and belongs in the blank. (**Note:** On the real GED test, the choices will appear as a "drop-down" menu. When you click on a choice, it will appear in the blank.)*

30. During meiosis, a parent cell with Select ... ▼ chromosomes divides

Select ... ▼
46
51
62

to produce daughter cells with Select ... ▼ chromosomes.

Select ... ▼
51
124
23

31. Compared to their parent cells, the cells resulting from mitosis each have Select ... ▼ DNA.

Select ... ▼
the same
less
more

POSTTEST

Question 32 refers to the following information.

An atom is the smallest possible particle of a substance and is made up of three particles: protons, electrons, and neutrons.

- Protons have a positive charge, "+".
- Electrons have a negative charge, "–".
- Neutrons do not have a charge at all. They are electrically neutral.

The atomic mass of an element is the total of the number of protons and neutrons. The **atomic number** is equal to the number of protons in the nucleus of an atom and determines the identity of the element.

32. Iron (Fe) has an atomic number of 26 and an atomic mass of 56. Which of the following correctly describes the nucleus of an iron atom?

    A. 26 protons, 30 neutrons
    B. 26 protons, 56 neutrons
    C. 56 protons, 56 neutrons
    D. 30 protons, 26 neutrons

POSTTEST

Questions 33 and 34 refer to the following information.

Physical properties—A physical property is an aspect of a substance that can be observed or measured without changing the identity of the substance.

Chemical properties—A chemical property is an aspect of a substance that can be observed only during a chemical reaction that alters the substance's chemical identity and produces one or more new substances.

Chemical and physical properties are thus related to chemical and physical changes of matter.

33. Indicate the correct box where each of the following belongs.

- Molecular weight
- Oxidation
- Temperature
- Melting point
- Momentum
- Acidity
- Rusting

A: Physical properties	B: Chemical properties

*The following question contains a blank marked "Select . . . ▼." Beneath it is a set of choices. Indicate the choice that is correct and belongs in the blank. (**Note**: On the real GED test, the choices will appear as a "drop-down" menu. When you click on a choice, it will appear in the blank.)*

34. Freezing of water into ice is an example of

Select . . . ▼
sublimation
evaporation
a physical change
a chemical change

.

POSTTEST

35. The following diagram represents the bright-spectra of four individual elements and the bright-line spectrum that is produced when three of the four elements are mixed together.

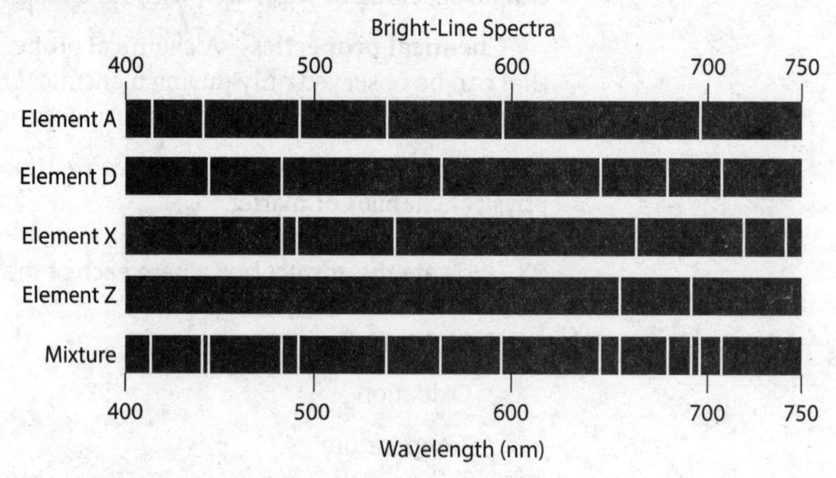

Which element is NOT present in the mixture?

A. element A
B. element D
C. element X
D. element Z

36. In humans, dark hair color, D, is dominant over light hair color, d. If the egg cell of a blonde female who carries only the genes for light-colored hair, dd, is fertilized by the sperm cell of a male who carries only the genes for dark hair, DD, what are the chances that the offspring will have dark hair?

A. 100 percent
B. 50 percent
C. 25 percent
D. 0 percent

POSTTEST

37. Mountains are formed when tectonic plates lift and tilt over as a result of grinding against each other. These gigantic movements and shifts cause collision and crumpling of Earth's crust and, in turn, form mountains. Which of the following is NOT a cause of mountain formation?

    A. tectonic plate shifts
    B. volcanic action
    C. global warming
    D. tectonic plate collisions

38. Mutation is a change in a chromosome that in some way alters the genetic message conveyed by that gene. Which of the following is a likely outcome of a mutation?

    A. alteration in the number of chromosomes of the gene
    B. change in the life span of the organism
    C. change in the physical appearance of the organism
    D. increased rate of cell division

POSTTEST

Question 39 refers to the following passage.

In a solution, two or more substances are joined in a homogeneous mixture with generally uniform physical properties. In such a mixture, a **solute** is a substance dissolved in another substance, known as a **solvent**. The ability of one substance to dissolve in another is called its **solubility**.

A basic principle of solution says that "like dissolves like." The overall solvation capacity of a solvent depends primarily on its polarity. Polar molecules are those that have an uneven distribution of electrons, whereas nonpolar molecules are those that have an even distribution of electrons. Polar/ionic solvents dissolve polar/ionic solutes and nonpolar solvents dissolve nonpolar solutes. For example, water is a polar solvent, and it will dissolve salts and other polar molecules, but not nonpolar molecules like oil. Petroleum is a nonpolar solvent and will dissolve in oil, but it will not mix with water.

*Questions 39 and 40 each contain one or more blanks marked "Select . . . ▼." Beneath each blank is a set of choices. Indicate the choice that is correct and belongs in the blank. (**Note:** On the real GED test, the choices will appear as a "drop-down" menu. When you click on a choice, it will appear in the blank.)*

39. The substance that is more likely to dissolve in water is Select . . . ▼.

Select . . . ▼
NaCl
solid I_2
O_2 gas
proteins

40. Tectonic plates vary in density. When pressure between an oceanic plate and a continental plate builds up, subduction can occur. When this occurs, the Select . . . ▼ plate is subducted because it is the Select . . . ▼ of the two plates.

Select . . . ▼
oceanic
continental

Select . . . ▼
lighter
denser

THIS IS THE END OF THE SCIENCE POSTTEST.
ANSWERS AND EXPLANATIONS BEGIN ON THE NEXT PAGE.

POSTTEST

Answers and Explanations

1. **C** Heat flows because of a difference in temperature. A warmer substance becomes cooler and a cooler substance becomes warmer when heat flows from one to the other.
2. **higher; lower** Heat flows from the substance with a higher temperature to the substance with a lower temperature.
3. **B** If one rotation of the sun at its equator takes 25 Earth days, then in an entire Earth year, the sun makes approximately 15 rotations. 365 days / 25 days = 14.6 rotations.
4. **eight planets** The sun is orbited by eight planets, five dwarf planets, and a huge number of comets and asteroids.
5. **respiratory; circulatory** Coordination between the heart and the lungs refers to the respiratory and circulatory systems.
6. **D** Liquid mercury is a better conductor of heat than glass is. Although both glass and mercury expand in contact with heat, the volumetric expansion of mercury is more than that of glass.
7. **A** Looking at the gel electrophoresis, you can see that the DNA from species A is almost the same as that from species B. Species C's DNA fragments produced a sequence that does not match either of the other two.
8. **C** This is a classic example of survival of the fittest. Both strains of bacteria were exposed to an antibiotic, and one lived on while the other died. There was no new food source for the bacteria because they were grown on the same petri dish. Also, there was no DNA change over four days to create resistance, a change that could only take place over many years.
9. **D** The partially digested proteins trigger the production of hydrochloric acid in the same direction. Hence, this is a positive feedback system.
10. **outer core** The liquid portion can be found in the outer core of Earth.
11. Graph A: isochoric. Pressure changes and volume remains constant, so work is not done.

    Graph B: isobaric. Pressure remains constant, but volume changes, so work is done.
12. **pressure; temperature; zero**
13. **conduction** Heat moves from the burner to the pot and to its handle.
14. **both convection and radiation** Thermal energy is transferred by radiation from the sun to the solar heater and then by convection from the solar heater to the person's body via the air.
15. **convection** Thermal energy moves up the chimney via the air.
16. **conduction** Heat is transferred by direct contact from the iron to the cloth.
17. **exothermic; released** This reaction releases heat in the form of energy; therefore, it is an exothermic reaction.
18. **B** An electrical toaster involves a conversion from electrical energy to thermal energy to toast the bread.
19. **B** Nuclear energy is obtained by nuclear fission, which is the process of splitting apart uranium atoms in a controlled manner that creates energy.
20. **D** The graph shows that the overshoot of population led to a decrease in the carrying capacity of the area.
21. **B** $v = \lambda \times f$, therefore,

    $v = (400 \times 2.5)$ m/s = 1,000 m/s
22. **B** According to the passage, gamma rays have the shortest wavelengths, then X-rays, ultraviolet waves, and visible light, which has the only electromagnetic waves we can see.

POSTTEST

Beyond that lie infrared and finally radio waves, which have the longest wavelengths.

23. **increases; increases** Larger numbers of tigers correlate with larger numbers of deer.
24. **directly proportional to** As the human population increases, so does the number of extinct species.
25. Epithelial tissue: skin. Nervous tissue: brain cells. Muscle tissue: biceps, cardiac muscle. Connective tissue: cartilage, ligament.
26. **A** Resultant force = (driving force − air resistance) = 3,600 N − 800 N = 2,800 N; mass of the car = 1,000 kg

    Per Newton's second law, force = mass × acceleration (a)

    2,800 N = 1,000 kg × a

    a = 2,800 N/1,000 kg = 2.8 m/s^2
27. **less than** When the water concentration inside is less, the water from outside will enter the cell.
28. Renewable sources of energy: solar, wind, geothermal, wood. Nonrenewable sources of energy: coal, oil.
29. **point B** Since the amniotic egg evolved in an early reptile, it is indicated as point B in the cladogram.
30. **46, 23** The number of chromosomes in daughter cells will be half of the number in parent cells.
31. **the same** DNA between parent and daughter cells is identical.
32. **A** The atomic mass (56) refers to the sum of the protons and neutrons in the atomic nucleus. The atomic number (26) refers to the number of protons in the nucleus. Therefore, the number of neutrons is (56 − 26) = 30.
33. Physical properties: momentum, temperature, melting point, molecular weight. Chemical properties: oxidation, acidity, rusting.
34. **a physical change** A physical change is reversible but a chemical change is not. The freezing of water into ice is a physical change because it can be reversed, as ice can be converted back to water on application of heat.
35. **C** According to the diagram, the spectral lines that appear in elements A, D, and Z all show up in the mixture. Note that the lines that appear for element X in the 650–750 nm range do not show up in the mixture. Therefore, element X is not in the mixture.
36. **A** Set up a Punnett square for this problem. A parent with genes for light hair (dd) is crossed with a parent with genes for dark hair (DD).

	d	d
D	Dd	Dd
D	Dd	Dd

Because all of the combinations contain a D, there is a 100 percent chance that all of the offspring will have dark hair.

37. **C** Mountains are formed primarily by tectonic plate collisions and other shifts, and by magma action from volcanoes.
38. **C** Mutation leads to change in the physical appearance of any organism.
39. **NaCl** It is a polar solute that dissolves in a polar solvent such as water.
40. **oceanic; denser** The denser plate will sink.

POSTTEST

Evaluation Chart

Circle the item number of each question that you missed. To the right of the item numbers, you will find the names of the chapters that cover the skills you need to solve the questions. More question numbers circled in any row means more attention is needed to sharpen those skills for the GED test.

Item Numbers	Chapter
5, 7, 8, 9, 20, 23, 24, 25, 29, 30, 31, 36, 38	Life Science
1, 2, 6, 11, 12, 13, 14, 15, 16, 17, 18, 19, 21, 22, 26, 27, 32, 33, 34, 35, 39	Physical Science
3, 4, 10, 28, 37, 40	Earth and Space Science

If you find you need instruction or more practice before you are ready to take the GED test, remember that we offer several excellent options:

McGraw-Hill Education Preparation for the GED Test: This book contains a complete test preparation program with intensive review and practice for the topics tested on the GED.

McGraw-Hill Education Pre-GED: This book is a beginner's guide for students who need to develop a solid foundation or refresh basic skills before they embark on formal preparation for the GED test.

McGraw-Hill Education Short Course for the GED: This book provides a concise review of all the essential topics on the GED, with numerous additional practice questions.